IS SCIENCE
NEUROTIC?

The Human World in the Physical Universe
(Rowman and Littlefield, 2001)

"Maxwell has not only succeeded in bringing together the various different subjects that make up the human world/physical universe problem in a single volume, he has done so in a comprehensive, lucid and, above all, readable way."

Dr. M. Iredale, *Trends in Cognitive Sciences*

"... a bald summary of this interesting and passionately-argued book does insufficient justice to the subtlety of many of the detailed arguments it contains."

Professor Bernard Harrison, *Mind*

"This is a very complex and rich book. Maxwell convincingly explains why we should and how we can overcome the 'unnatural' segregation of science and philosophy that is the legacy of analytic philosophy. His critique of standard empiricism and defence of aim-oriented empiricism are especially stimulating."

Professor Thomas Bittner, *Philosophical Books*

The Comprehensibility of the Universe
(Oxford University Press, 1998)

"Maxwell has clearly spent a lifetime thinking about these matters and passionately seeks a philosophical conception of science that will aid in the development of an intelligible physical worldview. He has much of interest to say about the development of physical thought since Newton. His comprehensive coverage and sophisticated treatment of basic problems within the philosophy of science make the book well worth studying for philosophers of science as well as for scientists interested in philosophical and methodological matters pertaining to science."

Professor Cory F. Juhl, *International Philosophical Quarterly*

"Some of [Maxwell's] insights are of everlasting importance to the philosophy of science, the fact that he stands on the shoulders of giants (Hume, Popper) notwithstanding .... Many of the pressing problems of the philosophy of science are discussed in a lively manner, controversial solutions are passionately defended and some new insights are provided; in particular the chapter on simplicity in physics deserves to be read by all philosophers of physics."

Dr. F. A. Muller, *Studies in History and Philosophy of Modern Physics*

From Knowledge to Wisdom
(Blackwell, 1984)

"This book is the work of an unashamed idealist; but it is none the worse for that. The author is a philosopher of science who holds the plain man's view that philosophy should be a guide to life, not just a cure for intellectual headaches. He believes, and argues with passion and conviction, that the abysmal failure of science to free society from poverty, hunger and fear is due to a fatal flaw in the accepted aim of scientific endeavour — the acquisition and extension of knowledge .... Maxwell is advocating nothing less than a revolution (based on reason, not on religious or Marxist doctrine) in our intellectual goals and methods of inquiry .... There are altogether too many symptoms of malaise in our science-based society for Nicholas Maxwell's diagnosis to be ignored."

Professor Christopher Longuet-Higgins, *Nature*

"The essential idea is really so simple, so transparently right .... It is a profound book, refreshingly unpretentious, and deserves to be read, refined and implemented."

Dr. Stewart Richards, *Annals of Science*

"... a strong effort is needed if one is to stand back and clearly state the objections to the whole enormous tangle of misconceptions which surround the notion of science today. Maxwell has made that effort in this powerful, profound and important book."

Dr. Mary Midgley, *University Quarterly*

"Maxwell has, I believe, written a very important book which will resonate in the years to come. For those who are not inextricably and cynically locked into the power and career structure of academia with its government-industrial-military connections, this is a book to read, think about, and act on."

Dr. Brian Easlea, *Journal of Applied Philosophy*

Is Science
Neurotic?

Nicholas Maxwell
University College London, UK

Imperial College Press

Published by

Imperial College Press
57 Shelton Street
Covent Garden
London WC2H 9HE

Distributed by

World Scientific Publishing Co. Pte. Ltd.
5 Toh Tuck Link, Singapore 596224
USA office: 27 Warren Street, Suite 401-402, Hackensack, NJ 07601
UK office: 57 Shelton Street, Covent Garden, London WC2H 9HE

British Library Cataloguing-in-Publication Data
A catalogue record for this book is available from the British Library.

ISBN   1-86094-500-7

Printed in Singapore.

For Chris

By the same author

———

What's Wrong with Science?
From Knowledge to Wisdom
The Comprehensibility of the Universe
The Human World in the Physical Universe

Faustus is Us

Faustus personifies our lust to know,
That science which we seek so avidly,
Since knowing leads to doing as we grow
In power to control our destiny.

But Faustus sold his soul for what he learned,
His bargain with the Devil granting him
A period of supremacy unearned
By merit, making his moral vision dim.

And so it is with us, so powerful
In realms of science and technology,
Who know so much of how, much less of why
Or what is wise, who rather play the fool
Than seek to earn the true supremacy
Of knowing how to live, not how to die.

Alan Nordstrom   21 March 2003

Preface

In this book I show that science suffers from a damaging but rarely noticed methodological disease, which I call rationalistic neurosis. It is not just the natural sciences which suffer from this condition. The contagion has spread to the social sciences, to philosophy, to the humanities more generally, and to education. The whole academic enterprise, indeed, suffers from versions of the disease.

It has extraordinarily damaging long-term consequences. For it has the effect of preventing us from developing traditions and institutions of learning rationally devoted to helping us learn how to make progress towards a wiser, more civilized world. On our fragile earth, overcrowded, fraught with injustice, inequality and conflict, menaced by unprecedented and terrifying technological and industrial means for change and destruction, we urgently need to acquire a little more wisdom and civilization if we are to avoid repeating horrors of the kind suffered by so many during the 20th century (and already suffered by many in the first few years of the 21st century). We can ill afford to have in our hands instruments of learning botched and bungled from the standpoint of helping us learn how to live more wisely.

It may seem scarcely credible that something as important as our institutions of learning can suffer from wholesale, structural defects. Why has this not been noticed before? Why have not armies of scientists and scholars appreciated the point and, long ago, put the matter right? The answer will emerge as my argument unfolds. I will show that one of the most damaging features of rationalistic neurosis is that it has built-in methodological and institutional mechanisms which effectively conceal that anything is wrong. But there is also an immediately obvious reason: specialization, and the resulting fragmentation of academia, has resulted in a situation in which hardly anyone takes responsibility for the overall ideals, the overall aims and methods, of academic inquiry. Academics, these days, are specialists, furiously trying to keep abreast of developments in their own specialist fields. They have no time, and no inducement, to lift their eyes from their particular disciplines and look at the whole endeavour.

Science has long been under attack, at least since the Romantic movement. Blake objected to "Single vision & Newton's sleep" and declared that "Art is the Tree of Life... Science is the Tree of Death". Keats lamented that science will "clip an Angel's wings" and "unweave a rainbow". Whereas the Enlightenment had valued science and reason as tools for the liberation of humanity, Romanticism found science and reason oppressive and destructive, and instead valued art, imagination, inspiration, individual genius, emotional and motivational honesty rather than careful attention to objective fact. Much subsequent opposition to science stems from, or echoes, the Romantic opposition of Blake, Wordsworth, Keats and many others. There is the movement Isaiah Berlin has described as the "Counter-Enlightenment" (Berlin, 1979, ch. 1). There is existentialism, with its denunciation of the tyranny of reason, its passionate affirmation of the value and centrality of irrationality in human life, from Dostoevsky, Kierkegaard and Nietzsche to Heidegger and Sartre (see, for example, Barrett, 1962). There is the attack on Enlightenment ideals concerning science and reason undertaken by the Frankfurt school, by postmodernists and others, from Horkheimer and Adorno to Lyotard, Foucault, Habermas, Derrida, MacIntyre and Rorty (see Gascardi, 1999). The soul-destroying consequences of valuing science and reason too highly is a persistent theme in literature: it is to be found in the works of writers such as D.H. Lawrence, Doris Lessing, Max Frisch, and Y. Zamyatin. There is persistent opposition to modern science and technology, and to scientific rationality, often associated with the Romantic wing of the green movement, and given expression in such popular books as Marcuse's *One Dimensional Man*, Roszak's *Where the Wasteland Ends*, Berman's *The Reenchantment of the World* and Appleyard's *Understanding the Present*. There is the feminist critique of science and conceptions of science: see, for example, Fox Keller (1984) and Harding (1986). And there are the corrosive implications of the so-called "strong programme" in the sociology of knowledge, and of the work of social constructivist historians of science, which depict scientific knowledge as a belief system alongside many other such conflicting systems, having no more right to claim to constitute knowledge of the truth than these rivals, the scientific view of the world being no more than an elaborate myth, a social construct (see Barnes and Bloor, 1981; Bloor, 1991; Barnes, Bloor and Henry, 1996; Shapin and Schaffer, 1985; Shapin, 1994; Pickering, 1984; Latour, 1987). This latter literature has provoked a counter-attack by scientists, historians and philosophers of science seeking to defend science and traditional conceptions of scientific rationality: see Gross and Levitt (1994), Gross, Levitt and Lewis (1996), and Koertge (1998).

This debate between critics and defenders of science came abruptly to public attention with the publication of Alan Sokal's brilliant hoax article 'Transgressing the boundaries' in a special issue of the cultural studies journal *Social Text* in 1996 entitled *Science Wars*: see Sokal and Bricmont (1998).

What does this book have to contribute, given the above avalanche of criticism of science? The criticisms of science developed in this book are *diametrically opposed* to the above, in so far as the above criticisms oppose scientific rationality, seek to diminish or restrict its influence, or hold that it is unattainable. My central point is that we suffer not from too much scientific rationality, but from not enough. What is generally taken to constitute scientific rationality is actually nothing of the kind. It is *rationalistic* neurosis, a characteristic, influential and damaging kind of *irrationality* masquerading as rationality. Science is damaged by being trapped within a widely upheld but severely defective philosophy of science; free science from this defective philosophy, provide it with a more intellectually rigorous philosophy, and it will flourish in both intellectual and humanitarian terms. And more generally, as we shall see, academic inquiry as a whole is damaged by being trapped within an intellectually defective philosophy of inquiry; free it from this defective philosophy, from its rationalistic neurosis, and it will flourish in intellectual and human terms. It is not reason that is damaging, but defective pretensions to reason — rationalistic neurosis — masquerading as reason.

Thus what I seek to do here is the exact opposite of what all those who oppose science and scientific rationality do. I shall argue that Reason, the authentic article, arrived at by generalizing the progress-achieving methods of science, can have profoundly liberating and enriching consequences for all worthwhile, problematic aspects of life, and thus deserves to enter into every aspect of life.

The bare bones of the argument of this book can be stated quite simply like this. Science cannot proceed without making the substantial metaphysical assumption that the universe is physically comprehensible (to some extent at least). But this conflicts with the orthodox view that in science everything is assessed impartially with respect to evidence, *nothing being permanently assumed independently of evidence*. So the metaphysical assumption of comprehensibility is repressed. Science pretends that no such assumption is made. But this damages science. For the assumption is substantial, influential and highly problematic. It needs to be made explicit within science so that it can be critically scrutinized, so that alternatives can be developed and considered. Pretending the assumption is not being made undermines the intellectual rigour of science, its intellectual value and success.

And it does not stop there. For science also makes value assumptions. Quite properly, science is concerned to discover that which is of value. New factual knowledge devoid of all value (whether intellectual or practical) does not contribute to science. But the orthodox view of science holds that values have no place within science. As in the case of metaphysics, here too, science pretends that values have no role to play within the intellectual domain of science. And this damages science. For values are, if anything, even more influential and problematic than metaphysics (in influencing the direction of research). Here too, values need to be made explicit within science so that they can be scrutinized, so that alternatives can be developed and assessed. Pretending that values play no role within the intellectual domain damages science; both the intellectual and the practical aspects of science are adversely affected. Science fails to pursue those avenues of research that lie in the best interests of humanity.

And it goes further. Science is pursued in a social, cultural, economic and political context. It is a part of various social, economic and political projects which seek to achieve diverse human objectives. But the idea that science is an integral part of humanitarian or political enterprises with political ends clashes, once again, with the official view of science that the aim of science is to improve factual knowledge. The political objectives of science are repressed. And, once again, this damages science. For, of course, the political objectives of science, like all our political objectives, are profoundly problematic. These need to be made explicit so that they can be scrutinized, so that alternatives can be developed and considered. The pretence that science does not have this political dimension once again undermines the intellectual rigour of science, and its human value. It lays science open to becoming a part of economic, corporate and political enterprises that are not in the best interests of humanity.

The upshot of the line of argument just indicated is that we need to bring about a revolution in the aims and methods of science, and of academic inquiry more generally. Natural science needs to change; its relationship with the rest of academic inquiry, with social inquiry and the humanities, needs to change; and, most importantly and dramatically, academic inquiry as a whole needs to change. The basic task of the academic enterprise needs to become to help humanity learn how to tackle its problems of living in more cooperatively rational ways than at present. We need to put the intellectual tasks of articulating our problems of living, and proposing and critically assessing possible solutions, possible and actual *actions*, at the heart of academic inquiry. The basic task needs to be to help humanity learn a bit more wisdom — wisdom being the capacity to realize what is of value in life, for oneself and others, wisdom including knowledge and technological know-how, but much else besides.

Natural science, despite its flaws, has massively increased our knowledge and technological know-how. This in turn has led to a massive and sometimes terrifying increase in our power to act. Often this unprecedented power to act is used for human good, as in medicine or agriculture. But it is also used to cause harm, whether unintentionally (initially at least) as when industrialization and modern agriculture lead to global warming, destruction of natural habitats and rapid extinction of species, or intentionally, as when the technology of war is used by governments and terrorists to maim and kill. Before the advent of modern science, when we lacked the means to do too much damage to ourselves and the planet, lack of wisdom did not matter too much. Now, with our unprecedented powers to act, bequeathed to us by science, lack of wisdom has become a menace. This is the crisis behind all the other current global crises: science without wisdom. In these circumstances, to continue to pursue knowledge and technological know-how *dissociated* from a more fundamental quest for wisdom can only deepen the crisis. As a matter of urgency, we need to free science and academia of their neuroses; we need to bring about a revolution in the academic enterprise so that the basic aim becomes to promote wisdom by intellectual and educational means. At present science and the humanities betray both reason and humanity.

The argument of this book, which I have just summarized, begins with a discussion of the philosophical and methodological problems that beset theoretical physics. At once, many of those who are concerned about moral, political and environmental issues that plague our world today, will feel impatient. What have the methodological problems of theoretical physics to do with third world poverty, war and the threat of war, pollution, extinction of species, the menace of conventional, chemical, biological and nuclear armaments?

I can only plead with such a reader: patience. The root cause of the sickness of our times does indeed lie, I claim, with a methodological sickness of natural science or, even more specifically, of physical science. What makes the modern world so utterly different from all previous ages is our possession of modern science and technology. This is the instrument that has changed the conditions of human life almost beyond all recognition, and put unprecedented, and sometimes terrifying, powers into our hands. One should not dismiss out of hand the suggestion that a part of our problem may lie with this instrument, this engine, of rapid change.

We see, here too, the way in which intellectual specialization, referred to above, has effectively concealed from view the nature and extent of the problem that confronts us. The argument that I shall develop in what follows begins with physics, with problems concerning the aims and methods of physics. But it then leaps to social science, to philosophy, to the humanities, to education, to psychotherapy, and to politics: in the end there is scarcely an aspect of modern life that is not touched and, potentially, affected by the argument. How many academic experts are prepared to follow, to take seriously, an argument that ranges so recklessly across such a wide range of academic disciplines? How many non-academic non-experts? But just this is what the argument and message of this book requires.

I hope that the reader will endure with patience the somewhat esoteric discussion concerning the aims and methods of physics pursued in chapter one. This may not seem to have anything much to do with the urgent problems of our times, but it does. This is in part how the intellectual disaster I seek to expose in this book conceals itself from view: it buries itself in the obscure, recondite field of the philosophy of physics. Those who devote their lives to the philosophy of physics are, by and large, too myopic to see how their specialized field of study has anything to do with the great and dreadful humanitarian problems and disasters of our age; and those who are above all concerned with these humanitarian disasters have no time at all for abstruse issues concerning the aims and methods of physics. And so the connection is never made.

As it happens, the methodological neurosis of physics, with which we begin in chapter one, is actually a quite simple matter to grasp. It does not require any real knowledge of physics to understand. Indeed, physicists and philosophers of physics are much more likely to find the argument difficult to follow than are non-experts. We non-scientists can stand back and see the whole wood; scientists, trained to think in terms of the current orthodox conception of science, trapped in the thickets of research, will find this much harder to do. In an attempt to take this peculiar circumstance into account, I have arranged the exposition as follows. In chapter one I give a simple exposition of the case for declaring natural science to suffer from what I call "rationalistic neurosis"; this avoids all technicalities, and goes straight to the heart of the matter. In the appendix I tackle a host of more technical objections that may be made to the elementary argument of chapter one. So, those experts in the fields of the philosophy of physics and philosophy of science who find the argument of chapter one naïve and unconvincing, should consult the appendix before tossing the book away in disgust.

I write in the hope that there will be a few who will not dismiss out of hand the suggestion that the question of how we are to go about learning how to live in wiser and more civilized ways might have something to learn from scientific learning, and will take the trouble to pursue the line of argument traced out in this book. I write in the hope that these few will grasp just how desperate our situation is, how urgent the need to change the *status quo*, and will do everything in their power to alert others to the need to heal the methodological sickness from which our institutions and traditions of learning at present suffer. To begin with, we need a campaigning organization, modelled perhaps on "Friends of the Earth", which might be called something like "Friends of Wisdom".

Nicholas Maxwell   July 2004

Acknowledgements

For helpful comments, suggestions, encouragement and criticisms made in connection with aspects of this book I would like to thank Chris van Meeteren, Leemon McHenry, Chris Isham, Gordon Fleming, F. A. Muller, Harvey Brown, Jos Uffink, Liz Stevenson, Paddy Clements, and my ever helpful Editor at Imperial College Press, Katie Lydon. I would like to thank Alan Nordstrom for his poem responding to my work, and for his permission to reproduce it here. This book is a very much extended, elaborated and rewritten version of an article with the same title published in *Metaphilosophy* (vol. 33, April 2002, pp. 259-299); I would like to thank the editor of that journal, Armen T. Marsoobian, for his best wishes for the book. My thanks to the Department of Education and Professional Development at University College London for financial assistance in connection with the production of this book.

Contents

Chapter One

The Natural Sciences

1.1 Rationalistic Neurosis

It seems, on the face of it, absurd to suggest that science is neurotic. Some scientists, along with other people, may be neurotic; even the odd pet. But how can a vast, impersonal intellectual endeavour like science be called neurotic? Is not this to attribute a mind to science, an ego, id and superego? What could be more nonsensical?

And even if it did somehow make sense to say of science that it is neurotic, wouldn't the assertion be patently false? Science has, after all, met with quite extraordinary success at improving our knowledge and understanding of the natural world. Could such an incredibly successful enterprise really be *neurotic*? If neurosis meets with such success should we not try to acquire it, rather than hope to be cured of it?

But let us consider a classic example of neurosis: the Oedipus complex. A boy loves his mother, and as a result is furiously jealous of, and hates, his father. But his father is big and powerful, and not easy to get rid of; and besides the boy also loves his father. So the hatred is repressed: see (Freud, 1962, pp. 77-78, 125-126). Nevertheless it persists into adult life, and one day, purely by accident, while caring lovingly for his elderly and ill father, the son mixes a lethal dose of medicine, and finally succeeds in fulfilling his long-repressed desire. But the act is rationalized away as a ghastly accident.

Put in a more abstract way, what one has here is something like the following. The son, whatever else he may be, is a being with aims, whether acknowledged or repressed. There is a basic desire or aim, A: to love his mother. There is a secondary, highly problematic, repressed aim, B: to kill his father. There is a third, declared, but somewhat unreal aim, C: to love, to care for, his father. The son supposes himself to be pursuing aim C while in reality he is pursuing aim B: actions performed in pursuit of B (administration of a lethal dose of medicine) are rationalized in terms of the pursuit of C (it was an accident): see Fig. 1.1.

The advantage of construing the Oedipus complex as a very special case of something much more general, namely the pursuit of problematic, repressed (or unacknowledged) aims under the smokescreen of apparently pursuing some unproblematic, acknowledged aim, is that it becomes possible to attribute neurosis to *anything* that can be construed (1) to pursue aims more or less successfully, (2) to represent (to itself or to others) the aims it pursues, and (3) almost inevitably, to *misrepresent* (some) aims that it is pursuing.

The aim-pursuing thing might be a person; or it might be an animal, a robot, a group of people, an institution, or a political, religious or cultural movement in so far as these can be construed to be aim-pursuing entities.

Neurosis, as I have sketchily characterized it above, is a condition that almost *any* aim-pursuing entity is likely to fall into, in so far as it is sufficiently sophisticated to represent, and hence misrepresent, the aims that it is pursuing. It is especially likely to arise when aims are problematic. Neurosis, conceived of in this way, is not a sickness of the psyche, the mind or the id; it does not require that there are mental acts of repression and rationalization; it does not presuppose, even, that the thing that suffers from neurosis is conscious or has a mind, not even in the sense that animals can be said to be conscious, or at least sentient. All that is required is that the thing in question pursues aims, represents the aims that it pursues, and hence on occasion misrepresents its aims. (At the very least we require that the thing in question can legitimately be *construed* to be aim-pursuing in this way.)

The notion of neurosis that I have indicated might be called "rationalistic neurosis" to distinguish it from Freudian or psycho-analytic notions. Rationalistic neurosis is a methodological notion, a notion that belongs to the theory of rational aim-pursuing.[1] It is especially damaging from the standpoint of rationality because, as the term "rationalization" implies, it subverts reason. Once a being has fallen into the pattern of confusion of rationalistic neurosis, "reason" becomes a hindrance instead of a help. The more "rationally" the being pursues its declared, false aim C, the worse off it is from the standpoint of pursuing its real, problematic aim B, the further away it is from *solving* the problems associated with the aim B, thus coming to pursue the really desirable aim A. The more "rationally" the being pursues its declared aim, the more unsuccessful, in real terms, it will be; in order to achieve real success the being must act "irrationally". Not only does this

---

1. The notion of rationalistic neurosis was first introduced by me in print in (Maxwell 1976a, pp. 206-221; 1984a, pp. 110-117).

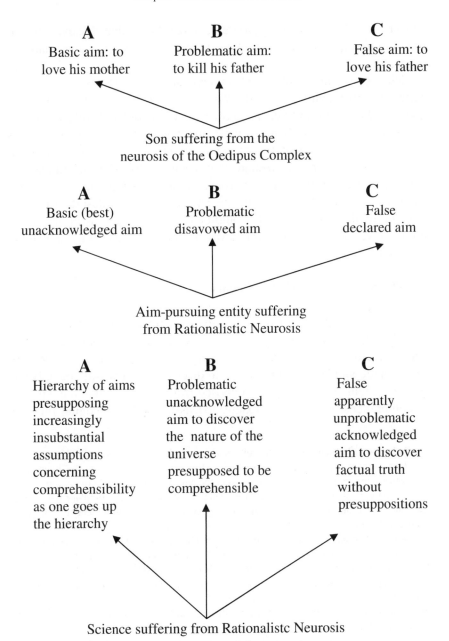

Figure 1.1: From the Oedipus Complex to the Neurosis of Science

subversion of reason block progress; it has the added disadvantage of bringing reason into disrepute. Reason appears to block, rather than aid, progress.

Science is an institutional endeavour that pursues aims; it is certainly sufficiently sophisticated to represent its aims, both to itself and to the public, in terms of its official "philosophy" (a philosophy of science being a view about what the aims and methods of science are, or ought to be). Thus, in terms of our new notion, it certainly makes sense to declare that science suffers from rationalistic neuroses. But is it true? I now proceed to demonstrate that it is. As we shall see, science has achieved its amazing success despite suffering from rationalistic neurosis. It would be even more successful, in a more humanly valuable way, if it threw off its current constricting, damaging neurosis.

1.2 The Neurosis of Natural Science

For science[2] to suffer from rationalistic neurosis, all that we require is that the real, problematic aim of science, B, differs from the official, declared aim, C. Just this is the case. A real aim of science, B, problematic and hence repressed, is to discover in what way the universe is comprehensible, it being presupposed from the outset that the universe is comprehensible (to some extent at least).[3] Acknowledging this aim

---

2. To begin with, "science" means "physics", even "theoretical physics". This may seem a rather narrow interpretation of science. But first, in discussing the neurosis of theoretical physics I am, in effect, discussing the neurosis of the whole of natural science, in that theoretical physics is, from an explanatory standpoint, the fundamental science, and all other branches of natural science presuppose physics. And second, I go on to consider broader neuroses of science that affect the whole of natural science, and technological research as well. And finally, in chapter three, I will consider even more serious neuroses of social science, and of academic inquiry when considered as a whole. These are all repercussions of the neurosis of theoretical physics with which we begin.

3. As I use the term here, to say that the universe is comprehensible is to say that it is such that there is *something* (God, society of gods, cosmic purpose, unified pattern of physical law), which exists everywhere, throughout all phenomena, in an unchanging form, and which, in some sense, determines or is responsible for all change and diversity, and in terms of which all change and diversity can, in principle, be explained and understood. If this *something* is a unified pattern of physical law, then the universe is *physically* comprehensible. If there is not just one *something* responsible for all change, but a number of distinct *somethings*, then the universe is only comprehensible *to some*

involves acknowledging that science accepts, from the outset as it were, as an article of faith, that the universe *is* comprehensible (to some extent at least).[4] But on what basis can this be known? To accept this substantial thesis about the nature of the universe as an article of faith makes science look more like a religion than what it is supposed to be; the sober, objective acquisition of reliable factual knowledge based on evidence. The aim is too problematic to be officially acknowledged, and hence is repressed, or disavowed.

Instead, the scientific community holds, officially as it were, that the basic intellectual aim of science, C, is to discover factual truths about the universe, nothing being permanently presupposed about the nature of the universe independently of evidence.[5] This declared, official aim seems

---

extent. The fewer the number, N, of distinct *somethings* that there are (other things being equal), so the more nearly perfectly comprehensible the universe is, perfect comprehensibility obtaining if $N = 1$. For further details see text, the appendix, section 2, and (Maxwell 1998, chs. 1, 3 and 4).

4. What we consider the aims of science to be may depend crucially on whether we are talking about "the context of discovery" or "the context of justification". By "the context of discovery" philosophers of science mean that context of scientific research in which new ideas, new theories, are being searched for and developed. "The context of justification", by contrast, is the context in which already formulated theories are assessed experimentally, to see whether they should be accepted to become a part of theoretical scientific knowledge, or rejected (see Reichenbach, 1961; see also Popper, 1959, p. 31). Everyone would agree that in "the context of discovery" all sorts of aims may be adopted; it is certainly permissible to adopt aims that presuppose that the universe is comprehensible. It is only in "the context of justification" that orthodoxy demands that the aim is to discover truth *nothing being permanently presupposed about what the truth is like.* In declaring that a basic aim of science is to discover in what way the universe is comprehensible, it being presupposed that the universe *is* comprehensible (to some extent at least), I am presupposing, here, the "context of justification", and not merely (and uncontroversially) "the context of discovery". In the text, "the aim of science" means "the aim of science in the context of justification".

5. By no means all scientists would agree that the official aim of science of discovering truth as such is the *real* aim of science. There are scientists who hold that an assumption about the simplicity or comprehensibility of nature is permanently implicit in the aim of science: Alan Sokal is one (personal communication). The mature Einstein might be cited as another (but even Einstein equivocates on the point: see (Maxwell 1993a, pp. 285-303) for a discussion. Despite this, we are justified, I claim, in holding that the scientific community as a whole holds, officially, that the aim of science (in the context of justification) is truth as such, no untestable (metaphysical) assumption being built

unproblematic; adopting it does not commit science to making some vast assumption about the nature of the universe, independently of the evidence. Adopting this aim enables scientists to hold on to the official view that the essential thing about science — that which distinguishes science from religions and other enterprises — is that in science claims to knowledge, laws and theories are accepted and rejected impartially on the basis of evidence, *no thesis about the nature of the universe being accepted permanently as a part of scientific knowledge independently of empirical considerations* (see Fig. 1.1). According to this view, considerations that have to do with simplicity, unity or explanatory power may influence choice of theory, in addition to empirical

---

permanently into the aim.   As evidence for this sociological thesis, I would cite the following. First, it is not hard to find scientists asserting that evidence alone, in the end, decides what theories are to be accepted and rejected. Here are two examples. Max Planck: "Experiments are the *only* means of knowledge at our disposal. The rest is poetry, imagination", quoted in (Atkins 1983, p. xiv). Poincaré: "Experiment is the sole source of truth. It alone can teach us something new; it alone can give us certainty" (Poincaré 1952, p. 140).   It is hard to find scientists asserting that some permanent metaphysical assumption *is* built into the aim of science.   More recently John Barrow has acknowledged that science makes presuppositions, but then declares that modern science has shown them to be false (Barrow, 1988, pp. 24-26). The point is echoed by Lewis Wolpert, who declares of Barrow's presuppositions "These assumptions may not be philosophically acceptable, but they are experimentally testable" (Wolpert, 1993, p. 107). Second, one may note that most scientists endorse some version of Popper's demarcation criterion between science and non-science, encapsulated in the remark that "...in science, *only* observation and experiment may decide upon the *acceptance and rejection* of scientific statements, including laws and theories" (Popper 1963, p. 54). This, again, is not compatible with the idea that untestable assumptions are built into the aim of science, and accepted as a part of scientific knowledge. Third, and most important, if it was officially acknowledged that untestable (metaphysical) assumptions are built into the aim of science, there would surely be official discussion in physics journals and textbooks as to what exactly these assumptions are, and on what grounds they are made. Metaphysics and epistemology would play some role in undergraduate physics courses. But this does not happen. This alone establishes decisively, in my view, that the official view of the scientific community is that while, in the context of discovery, one may hope that the universe is simple, or even comprehensible, in the context of justification no such assumption can be made. For further grounds for holding this sociological thesis, see: (Maxwell, 1984a, chs. 2 and 6; Maxwell, 1998, ch. 2). By "metaphysical assumption", here, I mean simply an assumption, such as "all events are caused", which is not empirically testable, not open to being refuted experimentally or observationally. "Epistemology" is the theory of knowledge, the study of how, to what extent, and in what sense, knowledge can be acquired.

considerations; this must not, however, commit science to making the permanent assumption that the universe itself is simple, unified or comprehensible.

But this official philosophy of science, which I shall call *standard empiricism*, taken for granted by scientists and non-scientists alike, is untenable.[6] Elementary considerations show that science cannot possibly have the specified aim, and cannot possibly proceed in the specified way. All versions of standard empiricism are untenable.

Given any scientific theory, however well verified empirically, there will always be infinitely many rival theories, equally well supported by the evidence, or even better supported by the evidence, which make different predictions, in an arbitrary way, for phenomena not yet observed.

Consider, for example, Newtonian theory, a theory that has met with extraordinary empirical success. It is easy to formulate endlessly many rival theories which are just as empirically successful as Newtonian theory, or even more successful empirically.

The nub of the theory is contained in two laws. There is Newton's famous law of gravitation: Given any two bodies, masses M_1 and M_2, distance d apart, then there is an attractive force of gravitation between the two bodies that is proportional to the product of the masses, and inversely proportional to the square of the distance between the two bodies (strictly speaking the distance between the centres of mass of the two bodies). In symbols, this can be written as:

(1) $F = GM_1M_2/d^2$, where F is the force of gravitation experienced by

---

6. Two versions of standard empiricism need to be distinguished. On the one hand there is "bare" standard empiricism which asserts that empirical considerations *alone* determine choice of theory in science (including also, possibly, considerations having to do with empirical content). This is defended in (Popper, 1959) and (van Fraassen, 1980). On the other hand there is "dressed" standard empiricism which asserts that two considerations govern choice of theory in science: those that have to do with empirical success, and those that have to do with the simplicity, unity or explanatory power of theories. The crucial point is that favouring simple theories must not commit science to assuming, permanently, that the universe itself is simple. Most scientists and philosophers of science uphold versions of "dressed" standard empiricism — see note 14 of the present chapter — even late Popper: see (Popper, 1963, p. 241)). Both versions are, I argue, untenable. It might seem that the fact that string theory is, so far, untestable goes against the claim that most scientists accept standard empiricism. This point is discussed in ch. 2.

each body, and G is a constant, Newton's gravitational constant.

In addition to (1), we require a law which specifies how an object moves when there is a force on it. This is given by Newton's force law, which asserts that a force F on a body of mass M has the effect of making that body accelerate; more precisely, the force F is equal to the mass M times the acceleration a, that is:

(2) $F = Ma$.

One striking feature of these two laws is that they apply to an extraordinarily wide range of phenomena; they predict the motions of objects in an extraordinarily wide range of circumstances. They apply to the moon as it goes round the earth, to the earth as it goes round the sun, and to the other planets and to comets; they apply to double star systems, to millions of stars rotating in galaxies, and to galaxies attracting each other across vast stretches of space; they apply to stones thrown on the earth's surface, and to molecules of hydrogen in outer space collapsing together to form dense clouds and eventually new stars. These two laws are put forward as applying to all massive objects everywhere, at all times and places, whatever the size, shape, constitution, temperature, colour, relative velocity, mass or distance apart the objects may be. The *explanatory power* of Newton's theory comes from this feature of the theory, namely that the *same* laws are put forward as applying to such a wide range of diverse phenomena.

We know now that Newton's laws do not predict the motion of objects accurately in all circumstances. If the relative velocity of objects is close to the velocity of light, or if gravitational forces are especially intense, then we need Einstein's more general and more accurate theory of gravitation, his general theory of relativity. If the massive objects are especially minute, for example fundamental particles such as electrons or protons, then we need quantum theory. But, as far as our present argument is concerned, we may ignore this complication. Let us treat Newtonian theory as a theory that has met with nothing but empirical success. We can, if we wish, imagine we are in the middle of the 19th century, and Newtonian theory has triumphed over all problems, and has successfully predicted all phenomena to which it has been applied.

Even in these circumstances, endlessly many rival theories can be formulated which are just as empirically successful as Newtonian theory, or which are even more successful empirically.

Thus, one rival theory might assert that everything occurs as Newtonian theory asserts up till midnight tonight when, abruptly, an inverse cube law of gravitation comes into operation. A second rival theory asserts that everything occurs as Newtonian theory predicts, except for the case of any two solid gold spheres, each having a mass of a thousand tons, moving in otherwise empty space up to a mile apart, in which case the spheres attract each other by means of an inverse cube law of gravitation. A third rival asserts that everything occurs as Newtonian theory predicts until thirty tons of gold dust and thirty tons of diamond dust are heated in a platinum flask to a temperature of 500°C, in which case gravitation will instantly become a repulsive force everywhere. There is no limit to the number of rivals to Newtonian theory that can be concocted in this way, each of which has all the predictive success of Newtonian theory as far as observed phenomena are concerned but which makes different predictions for some as yet unobserved phenomena.[7] Theories of this type can even be concocted which are *more* empirically successful than Newtonian theory, by adding onto Newton's theory independently testable and corroborated laws, or by arbitrarily modifying Newton's theory, in this entirely *ad hoc* [8] fashion, so that the new theory yields correct predictions where Newton's theory does not, as in the case of the orbit of Mercury for example (which very

---

7. A general procedure for concocting such rival theories can be indicated as follows. Any dynamical physical theory, such as Newtonian theory, can be regarded as specifying an abstract "space" (with a vast, perhaps infinite, number of dimensions), each point in the space corresponding to one specific kind of physical system to which the theory applies. In the case of Newtonian theory, there are points corresponding to systems consisting of two bodies, points corresponding to systems consisting of three bodies, and so on, with such and such masses and other features. However highly verified the theory is, the predictions of the theory will only have been verified for a minute region in the "space" of all possible phenomena predicted by the theory. In order to formulate a rival theory, all that one needs to do is specify some small region in the "space" of phenomena that has so far not been observed, and then substitute, for this region, any laws one pleases. The resulting *ad hoc* theory will have all the predictive success of the original theory. Endlessly many empirically successful, *ad hoc* rivals can be concocted in this way.

8. By an *ad hoc* theory, I mean simply a theory that postulates an abrupt, arbitrary change in its basic laws for some restricted range of phenomena, as already indicated in the text.

slightly conflicts with Newtonian theory).

One can set out to refute these rival theories by making the relevant observations or experiments, but as there are infinitely many of them, and each needs a different experiment to be refuted, this may take some time. In short, if science really did take seriously the idea that theories must be selected on the basis of evidence alone, science would be swamped by an infinity of empirically equally successful rival theories; science would come to an end.

And it would be the end of technology too. For, whenever well established scientific laws are used in connection with some industrial, engineering or medical process, however prosaic and standard, such as building a car or bridge, or manufacturing a drug, empirically more successful laws, concocted in the way indicated above, would predict utterly different outcomes: atomic explosions, collapsing bridges, drugs that are deadly poisons. Reliance on evidence alone would stymie science, stymie industry, and indeed all of human life.

None of this happens in practice because, in scientific practice, given an accepted, well verified theory, such as Newtonian theory, quantum theory, or general relativity, almost all the infinitely many equally empirically successful (and more successful) rival theories are, in comparison, grotesquely *ad hoc*, or disunified ("disunified" because these theories are, in effect, two or more distinct theories stuck arbitrarily together). Such theories are, in practice, excluded from scientific consideration on the grounds that they postulate an abrupt, arbitrary change in the laws for some restricted range of phenomena. The laws fail to be *invariant* as one moves, in imagination, through space and time, or from one range of phenomena to another. Such *ad hoc*, or disunified theories violate what physicists call "symmetry principles".

A symmetry principle in physics asserts that if a physical system is changed in some specific kind of way, it nevertheless continues to evolve in time as before. Thus, changing merely the location of a physical system in space does not change the way the system evolves: this is known as positional symmetry. Equally, changing merely the spatial orientation, velocity, or time of occurrence of a system does not change the way the system evolves. To each of these there corresponds a different symmetry (rotational symmetry, the symmetry of the restricted principle of relativity, temporal symmetry).

Empirically successful *ad hoc*, or disunified, rivals to accepted physical theories, of the kind indicated above, all violate symmetry principles of one kind or another. In particular, they all violate the general symmetry requirement that the *same*[9] laws should operate throughout the whole range of phenomena to which the theory applies.[10] These empirically successful rival theories are all rejected, or rather not even considered, not for empirical reasons, but because they violate these requirements of uniformity, unity, symmetry, and explanatory power.[11]

Most physicists (and even many philosophers of science) would agree with the argument so far. It is the next step which will provoke horrified disagreement.

For now comes the crucial point. In persistently excluding infinitely many such empirically successful but *ad hoc*, or disunified, theories, science in effect makes a big assumption about the nature of the universe, to the effect that it is such that no *ad hoc* (or disunified) theory is true, however empirically successful it may appear to be for a time. Without some such big assumption as this, the empirical method of science collapses. Science is drowned in an infinite ocean of empirically successful *ad hoc* theories.

If scientists only accepted theories that postulate atoms, and persistently rejected theories that postulate different basic physical entities, such as fields[12] — even though many field theories can easily

---

9. To say that the laws of a theory are "the same" throughout a range of phenomena seems, on the face of it, to have a clear enough meaning, especially when one takes into account the *ad hoc* rivals to Newton, whose laws clearly do *not* remain "the same". As it happens, a number of technical difficulties arise in connection with the assertion that "the same" laws govern a range of phenomena. These technical difficulties are tackled and solved in the appendix, section 2.

10. For a discussion of the role of symmetry principles in theoretical physics, and of the relationship between symmetry and theoretical unity, see the appendix, and (Maxwell, 1998, pp. 123-139 and 257-265).

11. In order to be explanatory, a physical theory must depict an underlying unity in ostensibly disparate phenomena. The theory must exhibit unity. This, in turn, is closely related to the content of the theory exhibiting symmetries, so that — at the very least — the theory postulates that the same laws govern the diverse phenomena to which the theory applies. For further details, see text below, the appendix, and (Maxwell, 1998, chs. 3 and 4, and pp. 257-265).

12. A "field" in theoretical physics is a (hypothetical) physical entity spread out

be, and have been, formulated which are even more empirically successful than the atomic theories — the implication would surely be quite clear. Scientists would in effect be assuming that the world is made up of atoms, all other possibilities being ruled out. The atomic assumption would be built into the way the scientific community accepts and rejects theories — built into the implicit *methods* of the community, methods which include: reject all theories that postulate entities other than atoms, whatever their empirical success might be. The scientific community would accept the assumption: the universe is such that no non-atomic theory is true.

Just the same holds for a scientific community which rejects all *ad hoc* (or disunified) rivals to accepted theories, even though these rivals would be even more empirically successful if they were considered. Such a community in effect makes the assumption: the universe is such that no *ad hoc* theory is true (unless implied by a true unified theory).

Thus the standard empiricist idea that science has the aim of improving knowledge of factual truth, *nothing being presupposed about the nature of the universe independently of evidence* is untenable. Science makes one big, persistent assumption about the universe, namely that it is such that no *ad hoc* theory is true. It assumes that the universe is such that there are no pockets of peculiarity, at specific times and places, or when specific conditions arise (gold spheres, gold and diamond dust, etc.), that lead to an abrupt change in laws that prevail elsewhere. Science assumes, in other words, that there is a kind of uniformity of physical laws throughout all phenomena, actual and possible. Furthermore, science *must* make this assumption (or some analogous assumption) if the empirical method of science is not to break down completely. The empirical method of science of assessing theories in the light of evidence can only work if those infinitely many empirically

---

continuously in space, which varies continuously in intensity from place to place and time to time, and which has the effect of exerting a force on an appropriately charged particle that happens to be in the field. (Conceivably, a field might be self-interacting, and thus exert a force on itself.) An example is the gravitational field surrounding the earth. This field exerts a force on all massive objects near the earth's surface. It causes stones, cups, and other objects to fall when dropped. Another example is the electromagnetic field postulated by classical electrodynamics. This exerts a force on electrically charged particles, and on magnets. The force varies in strength and direction as the field itself varies. The magnetic field surrounding a magnet can be made visible by covering the magnet with a piece of paper and sprinkling iron filings on the paper.

successful but *ad hoc*, disunified theories are permanently excluded from science independently of empirical considerations; to do this is just to make the big, permanent assumption about the nature of the universe.[13]

The academic discipline of the philosophy of science, in so far as it seeks to justify science in terms of the (declared, false) aim, C, of improving knowledge of truth, nothing being permanently presupposed about the truth, is engaged in a deeply neurotic activity. It is the neurotic face of science. It is providing, not reasons for the actions of scientists, but rationalizations.[14] The more nearly science conforms to the edicts of

---

13. For a much more detailed exposition of this refutation of standard empiricism see (Maxwell, 1998, ch. 2). See, too, the appendix.

14. The idea that science accepts theories that are sufficiently empirically successful, and sufficiently non-*ad hoc*, simple, unified or explanatory, *no permanent assumption being made by science about the nature of the universe*, is common ground to logical positivism (Ayer, 1936), inductivism (Carnap, 1950; Hesse, 1974), logical empiricism (Hempel, 1965), hypothetico-deductivism (Watkins, 1984; Miller, 1994), conventionalism (Duhem, 1954), constructive empiricism (van Fraassen, 1980), pragmatism (Peirce, 1958, chs. 5-11; Reichenbach, 1961) realism, structural realism (Worrall, 1989), induction-to-the-best-explanationism (Harman, 1965, 1968; Lipton, 2004), and the views of Popper (1959, 1963), Kuhn (1970), Lakatos (1970), and Holton (1973). For Bayesianism see the appendix, pp. 189-192. All these views, then, misrepresent the basic aim of science; they amount to rationalizations, rather than successful depictions of science as a rational enterprise. Views such as these constitute the neurotic face of science. Instead of contributing to the rationality of science, they serve to undermine it. Quine (1961) might seem to be a philosopher who has emphatically rejected standard empiricism, in that he has defended the view that logical and mathematical propositions are quasi factual or empirical in character, but central to knowledge and highly resistant to revision. But Quine's view is deeply flawed. The propositions of logic and pure mathematics are, despite what Quine says, sharply distinct from factual propositions, such as scientific theories. Logical and mathematical propositions, like "It's either raining, or it isn't", or "2 + 2 = 4" are interpreted and upheld in such a way that nothing that goes on in the world is permitted to falsify propositions such as these. (To say this is not to say that all logical and mathematical propositions can be known to be true with certainty.) Thus Quine's picture of logical and mathematical propositions being quasi-factual in character, not sharply distinct from obviously factual propositions but just more resistant to refutation, is untenable. In sharp contrast to Quine's picture, the assumption that the universe is such that no *ad hoc* or disunified theory is true is a massively *substantial* thesis about the nature of the universe, which might well be false, and which is not remotely like the propositions of logic or pure mathematics. John Dupré has declared recently that "It is now widely understood that science itself cannot progress without powerful assumptions about the world it is trying to investigate, without, that is to say, *a priori* metaphysics" (Dupré, 1993, p. 1). This sounds like the announcement of the

such philosophers of science, the more unsuccessful science becomes. (This accounts for the uselessness of much academic philosophy of science for science itself, a point sometimes made by working scientists, as we shall see below.)

In order to do justice to scientific practice we must acknowledge the real intellectual aim of science (aim B of Fig. 1.1): to improve knowledge about the universe *presupposed to be physically comprehensible*, to the extent at least that there is some yet-to-be-discovered, true, physical theory-of-everything, T, which is at least not grotesquely *ad hoc*, or disunified, like the *ad hoc* versions of Newtonian theory considered above.[15]

Granted this aim, the problem of how and why empirically successful *ad hoc* or disunified theories are to be excluded from scientific consideration disappears. Such theories clash with the basic presupposition that the universe is physically comprehensible, at least to the extent that it is not grotesquely *ad hoc*, and are to be dismissed on that account. The problem of excluding empirically successful disunified theories from science — in effect the problem of induction — turns out to be a typically *neurotic* problem. It only arises because the basic aim of science is misidentified. If the aim of science is misidentified as that of improving knowledge about the universe, nothing being presupposed about the nature of the universe, then all theories equally successful empirically must be treated equally, infinitely many *ad hoc* rival theories must be taken seriously, all theoretical knowledge disappears, and the full horror of the problem of induction[16]

---

demise of standard empiricism. But Dupré goes on to say that "empirical inquiry . . . provides the evidence on which such assumptions must ultimately rest" (Dupré, 1993, p. 2). Dupré's view, like that of Kuhn, Lakatos or Holton, is evidently a version of standard empiricism.

15. We require, in addition, perhaps, that the universe is presupposed to be such that the number of distinct kinds of fundamental physical entity, N, postulated by the true theory-of- everything, is not too large. If N is some enormous number, $10^{10^{10}}$ say, the universe can hardly be said to be physically comprehensible to any significant extent. What we should take the assertion that the universe is physically comprehensible to mean will be further clarified in the appendix, section 2.

16. The problem of induction is usually understood to be the problem of explaining how scientific theories can be *verified* by means of observation and experimentation, so that the theories acquire some degree of probability greater than zero. But the problem can be understood to be, more generally and fundamentally, the problem of understanding how

destroys science.   Identify the aim of science properly, and these neurotic horrors vanish.

But two big new problems emerge instead, namely:

(1) Granted that the intellectual aim of science presupposes that the universe is physically comprehensible (to some extent at least), what exactly ought this (untestable, metaphysical)[17] presupposition to be?

(2) What possible justification can there be for just accepting, as a basic, permanent part of scientific knowledge, that the universe is physically comprehensible, in the chosen sense?

And there are further problems:

(3) What does it *mean* to say that the universe is comprehensible, that the laws of nature exhibit unity? Above I suggested that it means that the same set of laws apply to all phenomena but, as we shall see in the appendix, various more or less technical objections can be made to this proposal. What it means to assert of a theory, or of the set of laws postulated by a theory, that it is *unified* is generally regarded as a major unsolved problem in the philosophy of science. Even Einstein confessed that he did not know how to solve the problem.[18]

(4) Granted that (3) can be solved, how are *degrees* of unity to be assessed? How can theories, or possible universes, be ordered with respect to the degree of unity they exhibit?

(5) Assuming   that   the   metaphysical   thesis   concerning   the

---

scientific theories can be *selected* by means of observation and experimentation (all questions of verification and knowledge being suspended), so that science is not overwhelmed by an infinity of equally viable theories.

17. As I indicated in note 5, metaphysical, as understood here, means simply not empirically testable, or falsifiable. (A theory is empirically falsifiable if a collection of statements about particular states of affairs, such as specifications of an experiment and the outcome, suffice to falsify the theory. In practice, auxiliary laws or theories are often required as well before a given theory becomes empirically falsifiable.)

18. Einstein confessed that he did not know how to give a precise definition of the "inner perfection" of a theory, by which he meant what I have called "unity": see Einstein (1949, p. 23). For a discussion, see Maxwell (1998, pp. 105-106).

comprehensibility or unity of nature, that is presupposed by science, cannot be justified, and must be accepted as a part of scientific knowledge as an article of faith, what becomes of the unique *scientific* status of science? How is science to be distinguished from religion? What entitles science to lay claim to being a superior source of knowledge? Are not the floodgates opened to every sort of religious fanatic, to fundamentalists and creationists of every description?

(6) On the other hand, if the thesis that the laws of nature exhibit unity really is a part of current scientific knowledge, so that science tells us that one and the same laws govern *all* phenomena, including those that occur inside our heads, what becomes of free will, of consciousness, of the meaning and value of human life? Are we not reduced to mere complex physical systems evolving in accordance with fixed, invariant physical laws?

This is a list of formidable problems indeed. No wonder the aim, B, of improving knowledge of the universe *presupposed to be comprehensible* is suppressed, or disavowed, by science. The official (false, neurotic) aim, C, of improving knowledge, *nothing being presupposed about the universe*, seems in comparison so utterly free of problems. But this is an illusion. As we have seen, it is impossible for science to pursue this ostensibly unproblematic aim honestly: that would lead to science being overwhelmed by infinitely many empirically successful disunified theories. Science would come to an end.

What actually happens is that science in practice pursues the problematic aim, B, but in a covert way, under the smokescreen of pursuing the officially declared aim, C. Just this, of course, is the pattern of confusion of rationalistic neurosis.

It might seem that this kind of methodological neurosis is not too damaging, as long as only token gestures are made towards pursuing the avowed, false aim, C. But even when neurosis is lightly worn in this fashion, it still has at least one damaging consequence. As a result of failing to acknowledge its real aim, B, science is unable to acknowledge the serious *problems* associated with this aim, and is thus unable to tackle these problems explicitly, as a part of scientific research. This, in turn, obstructs the task of solving these problems, and improving the problematic aim, B, as a result. Science throws away a rational method for the discovery of fundamental new theories, as we shall see in chapter two. For the truth is that, whereas the neurotic problem of induction is insoluble, the problems, (1) to (6), associated with the real aim of science, can be solved! But in order to solve them, in scientific practice, it is essential that science throws off its neurosis, acknowledges its real problematic aim, and starts to take seriously the task of solving the problems associated with this aim.

This is a particular illustration of the general point that freeing oneself, or an institution, from rationalistic neurosis, puts one in the position to acknowledge, to tackle, and perhaps to solve, the serious problems associated with one's real aims, problems which may well have led to the repression of the aim, and the neurosis, in the first place.

Let us, then, now tackle the above problems associated with the real, problematic aim of science.

1.3 Problems of the Real, Un-Neurotic Aim of Science

In what follows, I give as simple and non-technical an account of how the above six problems are to be solved as I can. A more detailed, technical discussion is given in the appendix. I take the six problems in turn.

(1) Granted that science must accept, as a part of scientific knowledge, the thesis that the universe is physically comprehensible to the extent, at least, that no seriously *ad hoc* or disunified theory is true, what precisely ought this thesis to be?

The first point to appreciate is that there is no single, sharp distinction between unity and disunity. (This is, in my view, a point of fundamental importance; it plays a crucial role in the argument to follow.) By "unity" we could mean merely that physical laws are the same throughout space and time. Or we could mean, in addition, that physical laws remain the same as other variables change, such as velocity, temperature, or mass (so that, for example, Newton's inverse square law of gravitation does not abruptly become an inverse cube law as masses of 1,000 tons are reached). Or, more restrictively still, we could mean (in addition) that there is only *one* force in nature, and not three or four distinct forces (such as gravitation, the electromagnetic force, and the weak and strong forces of nuclear physics). More restrictively still, we could mean that there is just one kind of particle in existence, or one kind of physical entity, a self-interacting field spread throughout space and time. Finally, and even more restrictively, we could mean that space, time, matter and force are all unified into one, unified, self-interacting entity.

Even more restrictive assumptions can be made, which specify the kind of entity or entities out of which everything is composed. And at the other end of the spectrum, much looser, less restrictive assumptions could be made which, if true, would still make science possible. Thus science could assume: the universe is such that local observable phenomena occur, most of the time, to a high degree of approximation, in accordance with some yet-to-be-discovered physical theory that is not too seriously *ad hoc* or disunified.

It is always possible, of course, that the universe only appears to be physically comprehensible (to some extent). Perhaps, as theoretical physics advances, everything will become increasingly complex (as even some physicists believe). Perhaps a malicious God is in charge, who has been controlling the universe up to now in such a way that it is as if physics prevails everywhere, but who, shortly, will startle us all by causing a series of dramatic, large-scale miracles to occur which violate all known laws. Perhaps as we probe deeper into physical reality we will discover that the universe exemplifies, not physical laws, but something that is closer to a computer programme (as some people have suggested). The universe may be comprehensible, but not *physically* comprehensible. That is, it may be that *something* exists — God, a society of gods, an overall cosmic purpose, a cosmic "computer" programme — which controls or determines the way events occur, and in terms of which, in principle, everything can be explained and understood: but this *something* may not be a unified physical entity, a unified pattern of physical law, and thus the universe, though comprehensible, is not *physically* comprehensible. Finally, the universe may not be comprehensible at all, and yet it might still be possible for us to live, and to acquire some knowledge of our local circumstances.

How do we choose between these endless possibilities? Science must make some kind of choice. It is all-important that science makes a good choice, since this choice will determine what (non-empirical) methods are employed by science to assess theories. If science chooses a cosmological thesis that is radically false, then science will only consider false theories, and will exclude from consideration all theories

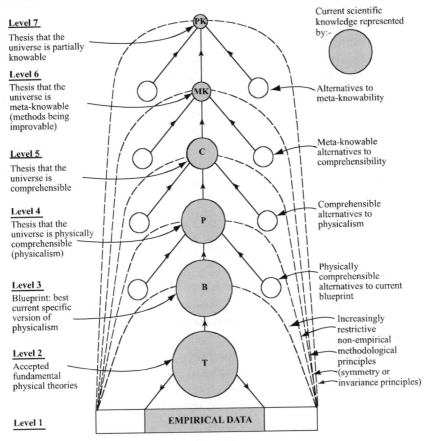

Level 7
Thesis that the universe is partially knowable

Level 6
Thesis that the universe is meta-knowable (methods being improvable)

Level 5
Thesis that the universe is comprehensible

Level 4
Thesis that the universe is physically comprehensible (physicalism)

Level 3
Blueprint: best current specific version of physicalism

Level 2
Accepted fundamental physical theories

Level 1

Current scientific knowledge represented by:-

Alternatives to meta-knowability

Meta-knowable alternatives to comprehensibility

Comprehensible alternatives to physicalism

Physically comprehensible alternatives to current blueprint

Increasingly restrictive non-empirical methodological principles (symmetry or invariance principles)

EMPIRICAL DATA

Figure 1.2: Aim-Oriented Empiricism

that might take one towards the truth. Science will come to a dead end. The more restrictive the chosen cosmological assumption is, so the more potentially helpful it will be in selecting theories, but also the more likely the assumption is to be radically false, thus imposing a block on scientific progress. On the other hand, the looser, the more unrestrictive the assumption is, so the more likely it is to be true, but the less helpful it will be in excluding empirically successful "disunified" theories. (Other things being equal, the less one says, the more likely it is that what one says is true. "The universe is not a chicken" is almost certainly true about ultimate reality, just because it says so little, there being an awful lot of ways in which the universe can not be a chicken.)

It is all-important that science makes the right assumption about the ultimate nature of reality; and yet it is just here, concerning the ultimate nature of reality, that we are most ignorant, and are almost bound to get things wrong. How on earth are we to proceed?

The solution to this dilemma — the fundamental dilemma concerning the real aim, B, of science — is to make not *one* cosmological assumption, but a *hierarchy* of assumptions, the assumptions becoming less and less restrictive, asserting less and less, as one goes up the hierarchy: see Fig. 1.2. At the top of the hierarchy there is a cosmological assumption that asserts so little it is very likely to be true, and such that its truth is necessary for knowledge to be possible at all. This is justifiably a permanent assumption of science. At the bottom of the hierarchy are assumptions so restrictive, so substantial in what they assert, that they are almost bound to be false. These assumptions will almost certainly need revision as science proceeds.

Figure 1.2 makes things look complicated, but the basic idea is extremely simple.[19] By displaying assumptions and associated methods — aims and methods — in this hierarchical fashion, we create a framework of relatively unspecific, unproblematic, fixed assumptions and methods (high up in the hierarchy) within which much more specific, problematic assumptions and methods (low down in the hierarchy) may be revised as science proceeds, in the light of the relative empirical success and failure of rival scientific research programmes to which rival assumptions lead. At each level, from 3 to 6, we adopt that assumption which (a) is compatible with the assumption above it in the hierarchy (in so far as this is possible), and (b) holds out the greatest hope for the growth of empirical knowledge, and seems best to support the growth of such knowledge (at levels 1 and 2). If currently adopted cosmological assumptions, and associated methods (represented by dotted lines in Fig. 1.2), fail to support the growth of empirical knowledge, or fail to do so as apparently successfully as rival assumptions and methods, then assumptions and associated methods are

---

19. An earlier and somewhat more complex version of this hierarchical view, with 10 instead of 7 levels, is to be found in (Maxwell, 1998); see also (Maxwell, 1999; 2001c, ch. 3 and appendix 3).

changed, at whatever level appears to be required.[20]   Every effort is made, however, to confine such revisions to cosmological theses as low down in the hierarchy of theses as possible. Only persistent, long-term, dramatic failure (at levels 1 and 2) would lead us to revise ideas above level 3, let alone above level 4; only an earthquake in our understanding of the universe would lead us to revise ideas above level 5. In this way we give ourselves the best hope of making progress, of acquiring authentic knowledge, while at the same time minimizing the chances of being taken up the garden path, or being stuck in a *cul de sac*. The hope is that as we increase our knowledge about the world we improve the cosmological assumptions implicit in our methods, and thus in turn improve our methods.   As a result of improving our knowledge we improve our knowledge about how to improve knowledge.   Science adapts its own nature to what it learns about the nature of the universe, thus increasing its capacity to make progress in knowledge about the world — the methodological key to the astonishing, accelerating, explosive growth of scientific knowledge.

This conception of science, postulating more or less specific, problematic, evolving aims and methods for science within a framework

---

20. How can level 3 assumptions, or assumptions higher up in the hierarchy, both influence, and be influenced by, level 2 theories? What makes this possible is a feature of the hierarchy about to be indicated in the text, namely that, as one goes up the hierarchy, assumptions are more and more firmly upheld. Level 2 theories that accord with the best available level 3 assumption tend to be favoured over rivals that do not so accord. Nevertheless, a level 2 theory that clashes with the current level 3 assumption, but (a) accords with the level 4 assumption, and (b) is more empirically successful than theories that are in accord with the best level 3 assumption, will be accepted, and will lead to the rejection, or modification, of the level 3 assumption with which it clashes. Consider, however, a theory that clashes, not just with level 3, but level 4 as well, even though compatible with level 5, in such a way that no version of the idea that the universe is physically comprehensible, at level 4, can be rendered compatible with the theory. Such a theory would have to meet with far greater, sustained empirical success before it led to the overthrow of the current level 4 assumption. It would have to lead to empirical research programmes across a broad front of natural science even more successful than current science, based on the current level 4 assumption, before it would become acceptable. And this would be the case even more, given a theory that clashes with the current level 5 assumption. In short, an assumption at a given level may, for much of the time, determine choices lower down in the hierarchy; but every now and again, it may itself be revised, because the revised version accords better with the assumption above, or is more empirically fruitful or, more likely, both of these simultaneously.

of more general, relatively unproblematic, more or less fixed aims and methods, I call *aim-oriented empiricism.*[21] The basic idea, let me re-emphasize, is that the fundamental aim of science of discovering how, and to what extent, the universe is comprehensible is deeply problematic; it is essential that we try to improve the aim, and associated methods, as we proceed, in the light of apparent success and failure. In order to do this in the best possible way we need to represent our aim at a number of levels, from the specific and problematic to the highly unspecific and unproblematic, thus creating a framework of fixed aims and meta-methods within which the (more or less specific, problematic) aims and methods of science may be progressively improved in the light of apparent empirical success and failure.

All this is a special case of a more general idea of *aim-oriented rationality* (to be discussed in chapter three), according to which, whenever basic aims are problematic (as they usually are in science and in life) we need to display aims at distinct levels of specificity and generality, thus creating a framework within which we have the best chance of improving more or less specific, problematic aims-and-methods as we proceed, in the light of success and failure.[22]

According to aim-oriented empiricism, then, scientific knowledge can be represented (in a highly schematic and simplifying way) as being made up of the following seven levels: see Fig. 1.2. At level 1, we have empirical data (low level experimental laws). At level 2, we have our best fundamental physical theories, currently general relativity and the so-called standard model (quantum theories of the fundamental particles and the forces between them). At level 3, we have the best currently available specific, but nevertheless metaphysical idea as to how the

---

21. Corresponding to each metaphysical, cosmological thesis, at level 3 to 7, there is a more or less problematic *aim* for theoretical physics: to specify that thesis as a true, precise, testable, experimentally confirmed "theory of everything". The aim corresponding to level 7 is relatively unproblematic: circumstances will never arise such that it would serve the interests of acquiring knowledge to revise this aim. As one descends the hierarchy of cosmological assumptions, the corresponding aims become increasingly problematic, increasingly likely to be unrealizable, just because the corresponding assumption becomes increasingly likely to be false. Whereas the uppermost aims and methods will not need revision, lower level aims and methods, especially those corresponding to level 3, will need to be revised as science advances. Thus lower level aims and methods evolve within the fixed framework of upper aims and methods.

22. For the generalization of aim-oriented empiricism to form aim-oriented rationality see also (Maxwell 1976a; 1984a, ch. 5; 1994a; 2001c, ch. 9).

universe is physically comprehensible. This asserts that everything is made of some specific kind of physical entity: corpuscle, point-particle, classical field, quantum field, convoluted spacetime, superstring, or whatever. Because the thesis at this level is so specific, it is almost bound to be false (even if the universe is physically comprehensible in some way or other). Here, ideas evolve with evolving knowledge, as a glance at the history of physics reveals. At level 4 we have the much less specific thesis that the universe is (perfectly) physically comprehensible in some way or other; and at level 5 we have the even less specific thesis that the universe is comprehensible in some way or other, whether physically or in some other way. At level 6 we have the thesis that the universe is such that there is some true, rationally discoverable thesis which, if accepted, enables us progressively to improve more specific assumptions and methods in the light of empirical success and failure, thus progressively improving methods for the improvement of knowledge. And at level 7 there is the thesis that the universe is such that we can acquire some knowledge of our local circumstances, sufficient at least to make life possible. This last thesis is so unspecific, so meagre, in what it requires of the universe for it to be partially knowable, that it can only help and can never hinder the pursuit of knowledge to accept it as a part of knowledge whatever the universe may be like. If it is false we cannot acquire knowledge whatever we assume, and however we proceed. It is thus justifiably a permanent part of scientific knowledge.[23]

It deserves to be noted that something like the hierarchy of metaphysical theses, constraining acceptance of physical theory from above, is to be found at the empirical level, constraining acceptance of theory from below. There are, at the lowest level of this empirical hierarchy, the results of experiments performed at specific times and places. Then, above these, there are low-level experimental laws, asserting that each experimental result is a repeatable effect. Next up, there are empirical laws such as Hooke's law, Ohm's law or the gas laws. Above these there are such physical laws as those of electrostatics or of thermodynamics. And above these there are theories which have been refuted, but which can be "derived", when appropriate limits are

---

23. For further details see the appendix; for details of the slightly more complex version of aim-oriented empiricism see (Maxwell 1998, ch. 1).

taken, from accepted fundamental theory — as Newtonian theory can be "derived" from general relativity. This empirical hierarchy, somewhat informal perhaps, exists in part for precisely the same epistemological and methodological reasons I have given for the hierarchical ordering of metaphysical theses: so that relatively contentless and secure theses (at the bottom of the empirical hierarchy) may be distinguished from more contentful and insecure theses (further up the empirical hierarchy) to facilitate pinpointing what needs to be revised, and how, should the need for revision arise. That such a hierarchy exists at the empirical level provides further support for my claim that we need to adopt such a hierarchy at the metaphysical level.

Let us turn now to the second big problem:

(2) What justification can there be for accepting, as a part of scientific knowledge, that the universe is comprehensible, whether this is understood in terms of the thesis at level 3, 4 or 5 (see Fig. 1.2)?

A crucial preliminary point must be made. Entirely in the absence of any justification whatsoever, the hierarchical conception of science of aim-oriented empiricism is more rational, more rigorous, than the orthodox view of standard empiricism. An elementary requirement for intellectual rigour is that the following principle is observed.

Principle of Intellectual Integrity: Assumptions that are substantial, influential, problematic and implicit need to be made explicit so that they can be critically assessed, so that alternatives can be developed and considered, in the hope that this may lead to the assumptions being improved.

The hierarchical view of aim-oriented empiricism does not just observe this principle: the view emerges, one might almost say, as a result of putting the principle repeatedly into practice. The hierarchy of cosmological assumptions concerning the comprehensibility and knowability of the universe makes explicit assumptions that are substantial, influential, problematic and implicit in the methods of science; furthermore, the hierarchy is designed to facilitate critical scrutiny, and improvement, of these assumptions, concentrating critical attention on those assumptions that seem most likely to need modification.

The neurotic, orthodox view of standard empiricism, by contrast, violates the principle of intellectual integrity: the attempt to pursue science in accordance with the edicts of standard empiricism leads to substantial, influential and problematic cosmological assumptions remaining implicit. The rigour of science is undermined.[24]

The next point to appreciate is that, in the end, all our knowledge is conjectural in character, as Karl Popper, for one, tirelessly emphasized (see Popper, 1959; 1963). The cosmological theses at levels 3 to 7 are all conjectures. It is especially important, in the interests of rigour, to appreciate that theses at levels 3 to 6 are conjectures, for only then do we appreciate that these theses may be false, and may need to be modified or replaced. The attempt to justify the truth of these conjectures, far from enhancing the rigour of science, does exactly the opposite, in that it tends to lead to complacency and dogmatism, to the view that none of these theses needs sustained critical scrutiny because their truth has been justified. Attempts at justification may block scientific progress, in other words.

This said, it must also be acknowledged that some conjectures are far

---

24. Thus aim-oriented empiricism is more rigorous than any version of standard empiricism (because the former accords with, and the latter violates, the principle of intellectual integrity) entirely independently of any justification of the metaphysical theses associated with aim-oriented empiricism. This is important for the following reason. The act of invoking metaphysical theses of "uniformity", like the theses of aim-oriented empiricism, is usually associated with the attempt to justify induction by an appeal to some such principle of "uniformity". Almost everyone recognizes that this is a hopelessly circular argument. Bas van Fraassen has described this approach in the following terms: "From Gravesande's axiom of the uniformity of nature in 1717 to Russell's postulates of human knowledge in 1948, this has been a mug's game" (van Fraassen, 1985, p. 260). But is not aim-oriented empiricism a version of this mug's game? The answer must be "No". If aim-oriented empiricism is more intellectually rigorous than standard empiricism (for the reasons given) quite independent of any attempt to solve the problem of induction then, when we do get round to trying to solve the problem of induction, we must begin with aim-oriented empiricism, and not with standard empiricism. It is absurd to try to solve the problem of induction, thus justifying scientific knowledge, in terms of a conception of scientific method (namely standard empiricism) which is demonstrably *unrigorous*, at the outset as it were. This leaves open, of course, how the vicious circularity of van Fraassen's "mug's game" is to be avoided, when we do get round to tackling the problem of induction granted aim-oriented empiricism. How this can be done is explained in section 6 of the appendix.

superior to others – in particular, those we hold to be a part of scientific knowledge. What justifies choosing the theses specified, at levels 3 to 7, in preference to rival theses, to be a part of scientific knowledge?

As I have already indicated, three considerations govern choice of cosmological theses at the different levels.

The top thesis, at level 7, is accepted as a permanent part of scientific knowledge on the grounds that, if false, acquisition of knowledge becomes impossible whatever is assumed. Accepting this assumption can only help, and cannot harm, the pursuit of knowledge whatever the universe may be like.

At levels 3 to 6, that cosmological thesis is accepted which (a) exemplifies (i.e. is an acceptable special case of) the thesis above (in so far as this is possible), and (b) holds out the greatest hope for the growth of empirical knowledge, and seems best to support the growth of such knowledge, at levels 1 and 2. Those theses are chosen, in other words, which seem to be the most fruitful from the standpoint of the growth of empirical knowledge.

This is the best that we can do. That a particular metaphysical view about the nature of the universe leads us to formulate more precise versions of this view, as testable theories which, again and again, meet with great empirical success, does not of course *prove* that the metaphysical view is correct. But it is reasonable to hold that the correct metaphysical view will be empirically fruitful, in this way. If we wake up in a pitch dark room, without the faintest idea where we are, and we make a guess which, again and again, leads us to make correct predictions about the position of furniture in the room, then it is reasonable to take this as grounds for holding that our guess is correct. This holds for science too. If our guess as to what kind of universe we are in leads to the development of theories which meet with great empirical success — more so than any rival guess put forward so far — then this provides grounds for accepting the guess as a part of our knowledge, at least until something better turns up.

The orthodox (neurotic) conception of science of standard empiricism holds that metaphysics needs to be eliminated from science (metaphysical ideas playing a role only in the context of discovery). But eliminating metaphysical assumptions from science is an intellectual disaster; it is this that creates the insoluble, neurotic problem of induction. Instead, we need to see science as employing a methodology which enables it to *improve* its metaphysical assumptions as it proceeds. It persistently seeks out (a) those that must be true if the acquisition of knowledge is to be possible at all, (b) those that offer the best hope of leading to the acquisition of knowledge, and (c) those that seem to have led to the greatest growth in empirical knowledge so far. This, to repeat, is the best that we can do.

But is the level 4 thesis that the universe is physically comprehensible (in the sense that the yet-to-be-discovered, true, physical theory-of-everything is perfectly *unified*), the most fruitful choice available? Is this thesis implicit in all successful theorizing in physics, from Galileo onwards? Is it even possible to see this degree of continuity in the history of theoretical physics? What of those great ruptures in the history of science, dramatized so effectively by Kuhn (1970) as scientific revolutions? Are not revolutions such that nothing theoretical survives through a revolution, as Kuhn maintains?

Revolutions, far from undermining the thesis that unity is the great, persistent assumption of theoretical physics, do just the opposite. All revolutions in theoretical physics, despite their diversity in other respects, reveal one common theme: they are all gigantic steps in unification. Thus Newton unifies Kepler's laws of planetary motion and Galileo's laws of terrestrial motion. Maxwell's theory of the electromagnetic field unifies electricity, magnetism and optics. Quantum theory unifies chemistry, properties of matter, and ultimately, with the development of quantum electrodynamics, electromagnetic phenomena. Einstein's theory of general relativity unifies special relativity, gravitation and the structure of space-time. Quantum electroweak theory, put forward by Weinberg and Salam, (partially) unifies the electromagnetic and weak forces. The so-called standard model (partially) unifies all known phenomena apart from gravitation. String theory, or M theory, if successful, will unify all phenomena. As I put it a few years ago: *"Far from obliterating the idea that there is a persisting theoretical idea in physics, revolutions do just the opposite in that they all themselves actually exemplify the persisting idea of underlying unity!"* (Maxwell, 1998, p. 181).[25]

---

25. John Dupré (1993) has argued that science, far from substantiating, has actually undermined, the thesis that there is unity in nature. This thesis will be contested in the appendix.

This aspect of theoretical physics, to which Kuhn fails entirely to do justice, is especially emphasized by aim-oriented empiricism. According to this view, revolutions in theoretical physics mark discontinuity at the level of theory, at level 2, and even discontinuity at level 3, but continuity at level 4. The level 4 thesis that underlying dynamic unity exists in nature, persists through revolutions — or, at least, has persisted through all revolutions in physics since Galileo. In order to make rational sense of physics (and natural science more generally), we need to interpret the whole enterprise as seeking to turn the level 4 assertion of underlying dynamic unity in nature, into a precise, unified, testable, physical "theory of everything". That, in a sentence, is what aim-oriented empiricism asserts. The thesis of underlying unity in nature, despite its metaphysical (untestable) character, is the most secure item of theoretical knowledge in science; it is by far *the* most fruitful idea that science has come up with, at that level in the hierarchy of assumptions. (An especially dramatic revolution would change even this level 4 thesis. This happened when Galileo rejected Aristotelian ideas, at level 4, and replaced them with the thesis that "the book of nature is written in the language of mathematics", which may be regarded as an early version of the current level 4 thesis of physical comprehensibility. In general, as I have already stressed, the higher up in the hierarchy discontinuity is introduced by a revolution, so the more dramatic, the more profound, the revolution will be, and the more extensive the changes will be to the nature of science, to the aims and methods of science.)

I consider, now, briefly, the remaining four problems associated with the real, repressed aim, B, of physics.

(3) What does it *mean* to assert that the universe is physically comprehensible, in the sense that the laws of nature exhibit *unity*?

The crucial requirement here, as I have indicated above, is that one and the same set of laws govern all phenomena, all the laws being required to predict the evolution of any given physical system. In assessing the unity of a theory, what matters is its content, what it asserts about the way phenomena evolve, and not its terminological or axiomatic form. I give a more detailed, technical discussion of this problem in the appendix, section 2.

(4) How are degrees to unity to be assessed?

Above, in the second paragraph of section 3, I indicated some ways in which "unity" can be interpreted in more and more restrictive senses: this provides a way of assessing degrees of unity. For further details see the appendix, section 2.

(5) If aim-oriented empiricism is accepted, how is science to be distinguished from religion?

The answer is that science accepts those metaphysical theses which either (a) are necessary for the pursuit of knowledge, or (b) seem to be the most fruitful from the standpoint of the growth of empirical knowledge. If the conjecture that an all-powerful, all-knowing, all-loving God were even more empirically fruitful, then it would, quite properly and rationally, be preferred to metaphysical theses currently accepted by science. If we lived in a world in which, whenever a child fell from a precipice, invisible hands swooped down from the heavens to save the child, and in general suffering and imminent death, caused by nature, were constantly averted by miraculous disruptions in physical law, we might well be rationally justified in adopting the conjecture that God exists and is in charge. But our actual world is very different.[26]

---

26. Put more pedantically, science consists of (a) reports of observational and experimental results (level 1 propositions); (b) testable theories (level 2); (c) metaphysical propositions concerning the comprehensibility and knowability of the universe which seem the most fruitful in promoting knowledge of type (a) and (b) (theses at levels 3 to 6); (d) metaphysical propositions accepted because knowledge can only be acquired if they are true (the thesis at level 7); (e) methodological principles governing choice of (level 2) theories (those specifying non-empirical considerations having to do with unity indicated by the slanting dotted lines of Fig. 1.2). Whether (a) to (e) includes any mention of God or not depends on how empirically fruitful a God-conjecture has proved to be. This aim-oriented empiricist picture of scientific knowledge will be enriched further when we come to discuss the role of values and politics in science in ensuing chapters. (Science also consists, quite essentially, of a vital dimension of know-how, of research skills in both experimental and theoretical science, the personal expertise and understanding of scientists, passed on informally from generation to generation, tacit knowledge as it has been called (see Polanyi, 1958), difficult if not impossible to make explicit as propositional knowledge.)

(6) If the universe really is physically comprehensible (as the level 4 thesis asserts), what becomes of free will, consciousness, the meaning and value of life?

This is a huge problem which cannot be tackled in this book. It can, however, in my view, be solved: see my book *The Human World in the Physical Universe* (2001c). A crucial point to appreciate is that physics is concerned only with a highly selected aspect of all that there is, that aspect which determines (perhaps probabilistically) the way events unfold. In addition to the physical, there are the experiential and value aspects of things: colours, sounds and smells as we experience them; inner experiences, feelings, thoughts, desires, intentions, and states of awareness; the meaning of utterances and writing; and value-features of people, other sentient creatures, and works of art. These experiential and value features of things can be explained and understood empathically or "personalistically" — a mode of understanding that is compatible with, but not reducible to physics. The upshot is that it is possible to understand how such things as free will, consciousness, meaning and value can exist even though the universe is physically comprehensible.[27]

I conclude that the problems associated with the real aim, B, of science are not insuperable. But in order to be solved in scientific practice they need to be acknowledged and confronted, and not, as at present, disavowed.

In so far as the scientific community continues to hold that the basic intellectual aim of science is to improve knowledge about the world, nothing being permanently presupposed about the world independently of evidence, all claims to knowledge in science being assessed impartially on the basis of empirical success and failure, science will continue to suffer from the rationalistic neurosis diagnosed above.[28] But

---

27. For details see (Maxwell, 2001c, especially chs. 2, and 5-9). The first sentence of the book formulates the problem (the rest of the book sets out to solve) like this: "How can we understand our human world, embedded as it is within the physical universe, in such a way that justice is done both to the richness, meaning, and value of human life on the one hand, and to what modern science tells us about the physical universe on the other hand?"

28. Recently, scientists and philosophers of science have leapt to the defence of the rationality of science in the light of postmodernist and sociological attacks: see (Sokal and Bricmont, 1998; Koertge, 1998). What both sides of this battle overlook, as I tried to explain in the preface, is that science as currently pursued and understood is not rational

when the scientific community abandons this official conception of the aims and methods of science and conceives of science more or less along the lines of aim-oriented empiricism, as just indicated, and furthermore, quite explicitly puts this conception into scientific practice, science will have freed itself from its neurosis.

At least three steps must be taken if science is to free itself of the rationalistic neurosis diagnosed above. First, the scientific community needs to recognize that all versions of standard empiricism are untenable, the real aim of science being, not just to improve knowledge of the universe, no presupposition being made about its nature but, on the contrary, to improve knowledge of the universe *presupposed to be more or less physically comprehensible.* This first step involves rejecting the false, official aim, C (of Fig. 1.1) and recognizing the problematic real aim, B. The second step involves recognizing that the *problems* associated with B need to be tackled *as an integral part of scientific research itself.* This step involves recognizing that untestable, metaphysical theses about the world are a part of scientific knowledge: Popper's criterion of demarcation has to be rejected.[29] It also involves appreciating that the philosophy of physics, properly conducted as the invention and criticism of rival possible views about what ought to be the aims and methods of physics, is a vital part of physics itself, influencing and being influenced by, progress in physics itself. The third step involves putting aim-oriented empiricism into practice, so that the more specific and problematic aims and methods of physics may be improved in the light of improving scientific knowledge, within a framework of relatively fixed and unproblematic aims and methods. The outcome is a conception of science, and of scientific method, dramatically different from what we have at present.

It is interesting to note, however, that the neurosis of science, being built into the intellectual-institutional structure of science, possesses a built-in method for protecting itself against its destruction. According to standard empiricism, an idea, in order to be a potential contribution to science, must be empirically testable.[30] But a diagnosis of the

---

enough.

29. For Popper, a theory is only scientific if it is empirically falsifiable; if it is not falsifiable, it is metaphysical and unscientific: see (Popper 1959, pp. 40-42).

30. This is Popper's falsifiability criterion of demarcation: see previous note.

methodological or rationalistic neurosis of science is not itself a straightforward testable contribution to factual knowledge. Hence, it will be excluded from science. Standard empiricism, once accepted by the scientific community and built into the institutional constitution of science, excludes such criticism of itself, on the grounds that such criticism is "philosophy of science", not science. The argument of this book would not qualify for publication in a respectable science journal. The neurosis of science has its own mechanisms of defence, in other words, a feature of the situation that mirrors methodologically points noted by Freud, Jung and other analysts of the human psyche.[31]

1.4 Does the Neurosis of Science Matter?

As I remarked at the outset, science is one of the most astonishingly successful human endeavours ever. It has improved knowledge and understanding in leaps and bounds — almost, one is inclined to think, at an ever accelerating rate. Without science, the modern world is inconceivable. Given this amazing success, does the neurosis of science really matter? If such success can be achieved by a neurotic enterprise, is not neurosis a boon rather than a hindrance?

How damaging a neurosis is depends on how seriously the false, avowed aim and associated methods are taken. If they are taken very seriously indeed, much energy and activity being devoted to the attempt to pursue the avowed aim, and to rationalize actions as being designed to realize this aim, then the neurosis will be very damaging. But if only lip service is paid to the false, avowed aim, and the real, problematic aim is, in practice, pursued resourcefully and intelligently, then the neurosis will not matter too much. As I have already suggested, this, fortunately, is how it is with science, up to a point at least. The neurosis does not bite too deep. Scientists pay only lip service to the false, avowed aim of science; in their research they take it for granted that explanations for phenomena exist to be found, the universe being (more or less)

---

31. I am personally well aware of the resistance of neurotic science to analysis, having been engaged in trying to get the argument of this book, in one form or another, across to the scientific and academic community for at least a quarter of a century: see (Maxwell, 1974) for an early attempt; see also (Maxwell, 1976a, 1977, 1979, 1980, 1984b, 1985a, 1987, 1991, 1992a, 1992b, 1993a, 1994a, 1997a, 1997b, 1998, 1999, 2000a, 2000b, 2001a, 2001b, 2001c, 2002, 2003a, 2003b, 2003c, 2004a, 2004b, 2004d, 2004f, and especially 1984a).

comprehensible, non-explanatory hypotheses and theories being ignored. Science has made such extraordinary progress despite, and not because of, the officially accepted philosophy of science of standard empiricism.

This said, it must also be emphasized that the neurosis of science does have a number of damaging consequences. Or, equivalently, freeing science of its neurosis would have a number of good consequences. In chapters two and three I indicate what some of these consequences are.

Chapter Two

Implications for Natural Science

2.1 Rational Scientific Discovery

Even though science has been extraordinarily successful despite its rationalistic neurosis, despite accepting the official, untenable, unworkable philosophy of science of standard empiricism (which misidentifies the basic aim and methods of science), it would be even more successful if neurosis was overthrown, standard empiricism was repudiated, and aim-oriented empiricism was adopted and put explicitly into scientific practice instead.

Granted standard empiricism, it is a mystery as to how new fundamental physical theories are discovered. If such discovery involved extending existing theories, the thing might not be such a mystery. But new theories almost always contradict earlier theories. Newton's theory of gravitation contradicts Kepler's laws of planetary motion and Galileo's laws of terrestrial motion; Einstein's theory of gravitation contradicts Newton's. Quantum theory contradicts the whole of classical physics; and relativistic quantum theory contradicts non-relativistic quantum theory: see (Maxwell, 1998, pp. 124-125, 211-217) for details. How, then, are these new theories discovered?

Standard empiricism cannot answer this question; and those who defend versions of standard empiricism, such as Popper, tend to hold that there is no rational method for the discovery of new theories in physics: see (Popper, 1959, pp. 31-32).

But granted aim-oriented empiricism, the situation is very different. Whereas, given standard empiricism, scientific knowledge consists of just empirical data and testable laws and theories (levels 1 and 2 of Fig. 1.2), given aim-oriented empiricism, scientific knowledge consists of items at *seven* distinct levels. In particular, at level 3 there is the best current untestable metaphysical conjecture as to how the universe is physically comprehensible; at level 4 there is the somewhat vaguer metaphysical thesis that the universe is physically comprehensible in some way or other; and at level 5 there is the even vaguer thesis that the universe is comprehensible (whether physically or in some other way).

An important point is that the level 4 thesis clashes with theoretical knowledge at level 2. The level 4 thesis asserts that the universe has a *unified* dynamic structure (some yet-to-be-discovered *unified* theory of everything being true) whereas, because of our ignorance, theories at level 2 clash with one another, and fail to form a unified whole. The conjecture at level 3 is the best available compromise between the vaguely asserted *unity* at level 4 and the precisely asserted *disunity* at level 2. This compromise will clash with the disunity of level 2, and will no doubt clash with the unity of level 4.

In seeking to discover better level 2 theories, physicists now have quite definite tasks to perform. Granted that there are two accepted fundamental physical theories which clash, everything inessential to the clash needs to be pared away, until two clashing ideas are arrived at, kernel ideas from each theory, in an attempt to get at the root of the clash. Taking clues in this way from clashing existing fundamental theories at level 2, and the unity at level 4, the basic task is to discover how to modify the thesis at level 3 so that it is a better compromise between levels 2 and 4; any modified thesis that emerges then needs to be made more and more precise until it becomes a new testable theory. In putting forward modified versions of the level 3 conjecture, physicists will, in effect, be developing new physical principles, new symmetry principles, which may act as guidelines for the construction of new theories. As I remarked in the last chapter, as a result of acknowledging the real and problematic aim of science, scientists are able explicitly to tackle the problems associated with this aim, this leading, perhaps, to an improvement in the aim.

It was in this way that Einstein discovered special and general relativity (Maxwell, 1993a, pp. 275-296). And, following Einstein's lead, it was in this way that more recent fundamental physical theories have been discovered, in particular the locally gauge invariant theories of quantum electroweak dynamics (which partially unifies electromagnetism and the weak force) and quantum chromodynamics (the theory of the strong force): see (Maxwell, 1998, pp. 135-139, and further references cited there).

The rational, but non-mechanical and fallible method of discovery just indicated is impossible granted standard empiricism. It requires that untestable, metaphysical ideas are rationally assessed, in terms of the justice they do to (1) unity and (2) accepted physical theories. But according to standard empiricism the only rational way of assessing ideas in science is in terms of empirical success and failure. One may, perhaps, by extension, assess untestable, metaphysical ideas in terms of their compatibility with existing accepted theories: but this is precisely the *wrong* way to assess ideas, granted that we seek to discover new theories. We require ideas that *clash* with existing theories. Thus, in so far as standard empiricism gives any guidelines for constructing new theories, they point in the wrong direction.

In some respects, current theoretical physics proceeds in a way which is much closer to aim-oriented than to standard empiricism. Theoretical physicists have invested a massive amount of work in developing string theory, or M-theory as it is now known. And yet, so far, no successful predictions have been forthcoming.[1] Given standard empiricism, this is wildly unscientific behaviour.[2] Given aim-oriented empiricism, string theory, or M-theory is an attempt to develop new level 3 ideas, and is thus entirely scientifically respectable. Furthermore, string theory seeks to unify the current two fundamental, clashing theories of physics, general relativity and quantum theory (or the current quantum field theory of particles and the forces between them, the so-called standard model). To this extent, current work on string theory proceeds in accordance with the method of discovery of aim-oriented empiricism.

In two other respects, however, work on string theory reveals the influence of standard empiricism, and fails to accord with aim-oriented empiricism.

The first point has to do with the way string theory was discovered and developed. String theory emerged, and was developed, as a result of the effort to create new mathematical tools for calculating results (see appendix, pp. 197-198). It did not seek, and has not altogether resulted in, a new unifying idea which resolves the clash between Einstein's theory of general relativity, and the quantum theory of particles and fields. This would involve unifying spacetime on the one hand, and

---

1. For a fascinating informal exposition and criticism of string theory by experts see (Davies and Brown, 1988). For an excellent, non-technical, more recent account see (Greene 1999); chapter 9 discusses the question of empirical predictions of the theory.

2. Many physicists object to work on string theory precisely because of its failure to yield experimental predictions, and consequent unscientific character (see for example the contributions of Sheldon Glashow, pp. 180-191, and Richard Feynman, pp. 192-210, in Davies and Brown, 1988); in objecting in this way, these physicists reveal their allegiance to standard empiricism.

matter (or particles-and-fields) on the other. String theory (or M-theory) does not (yet) achieve this, as string theorists themselves recognize: see (Greene, 1999, pp. 374-380).

The second point is this: because standard empiricism is still the official philosophy of science among physicists, there is a persistent lack of understanding as to how untestable metaphysical theories, such as string theory, are to be rationally developed and assessed. There is a certain failure among string theorists, for example, to appreciate the importance of trying to develop a number of rival level 3 ideas;[3] and there is a widespread failure to appreciate that such work can be *rationally* (if fallibly) assessed even before it issues in empirical predictions. Because of the failure to appreciate that work of this type can be assessed rationally, in practice what tends to influence this work is mere *fashion*. The vast majority of theoretical physicists working in this field of quantum gravity work on string theory (or M-theory), the fashionable thing to do; relatively few physicists explore other lines of inquiry. Few indeed are the physicists attempting to assess rationally the relative merits, the relative successes and failures of the rival research programmes.[4] Thus, even though standard empiricism is quite blatantly violated in scientific practice, its influence still lingers on and prevents aim-oriented empiricism from being put explicitly and fully into practice.

One outcome of rejecting standard empiricism and putting aim-oriented empiricism into scientific practice instead would be that physics would tend not to experience revolutions of the kind so brilliantly described by Kuhn (1970). This is because aim-oriented empiricist physics contains a rational, if fallible, method for the discovery of revolutionary new theories. As we have just seen, this involves the development and exploration of rival metaphysical ideas, at level 3,

---

3. This point is made in (Smolin, 2000, ch. 13). In particular, Smolin describes the way string theorists, and those pursuing the rather different approach to unifying quantum theory and general relativity called "loop quantum gravity", have tended to ignore, and even unjustly disparage, each others' work.

4. The most notable physicist known to me who does attempt to do this is Chris Isham: see (Isham, 1997). Smolin, too, seeks to bring together different approaches to quantum gravity (the hoped-for theory that unifies the quantum field theories of the standard model and general relativity).

which do better justice to the level 4 thesis that the universe has a unified dynamic structure. Such metaphysical ideas which promise to lead to greater theoretical unity, at level 2, are then developed until new empirically testable theories emerge. A new, revolutionary (level 2) theory, that dramatically contradicts earlier theories, arises, in other words, as a result of a long process of explicit scientific discussion of metaphysical ideas. There is no breakdown of rationality, of the kind depicted by Kuhn (1970, chs. 9 and 10). Ideas change at levels 2 and 3, but as a result of rational scientific discussion; and ideas do not change at the more fundamental level 4 (or, perhaps, level 5). The abandonment of a whole world view, and the acceptance of an entirely new world view, which Kuhn depicts as being an essential feature of a revolution (see Kuhn, 1970, ch. 10), does not occur.

In sharp contrast to all this, the attempt to pursue science in accordance with standard empiricism represses the explicit discussion and rational assessment of metaphysical ideas that lead up to the new, revolutionary theory. The result is that the new theory — the new paradigm, as Kuhn called it — bursts upon the scientific scene abruptly, without prior, sustained discussion of metaphysical ideas. Furthermore, granted standard empiricism, the new theory can only be assessed *empirically*. This, as Kuhn stresses, is likely to be inconclusive, at least in the short term; hence the (apparent) breakdown in scientific rationality during scientific revolutions. In addition, standard empiricism demands that the one strand of theoretical scientific knowledge that persists through a revolution — namely the level 4 metaphysical thesis that the universe has an underlying unified dynamic structure — is repressed, precisely because it is metaphysical. Hence, it can seem, as Kuhn stresses, that a revolution involves a complete change in world view. In short, the Kuhnian phenomenon of scientific revolutions are a symptom of the neurosis of science. Kuhnian revolutions occur (in so far as they do occur) because the scientific community attempts to pursue science in accordance with standard empiricism. Free science of its neurosis, and Kuhnian revolutions will tend to disappear. In so far as new, revolutionary theories emerge on the scientific scene, they will tend to emerge as a result of sustained, prior, rational discussion of metaphysical ideas, and will not burst upon the scene abruptly, in response merely to the accumulation of empirical anomalies and growing crisis, in the way depicted by Kuhn.[5]   Whereas aim-oriented empiricism facilitates the development of good candidates for radical new theories, and good ideas for such theories, standard empiricism tends to block this by outlawing the kind of innovative metaphysical thinking that is required from the domain of explicit scientific discussion. Kuhnian revolutions are symptomatic of a measure of irrationality induced by the attempt to put standard empiricism into scientific practice.

2.2 The Philosophy of Science

Freeing science of its neurosis would bring about a major and much needed revolution in the philosophy of science. At present most work in the philosophy of science, in the academic discipline that is, proceeds within the framework of, and seeks to justify standard empiricism (of one version or another): see (Maxwell, 1998, ch. 2). But this, as we have seen, is exactly the wrong thing to do. Current philosophy of science is a deeply neurotic activity. Not only is the philosophy of science beset by long-standing problems about the nature of science, which resist all attempts at solution — most notably problems of induction and simplicity.[6] In addition, work done in the philosophy of

---

5. To some extent, this is already a feature of theoretical physics. The revolutions brought about by Einstein, in creating special and general relativity, came about as a result of Einstein's search for underlying theoretical unity; this was of far greater importance than empirical problems facing Newtonian mechanics. Something similar can be said, to a lesser extent, concerning Einstein's important contributions to the development of quantum theory. (For a discussion of these issues see Maxwell, 1993a, 275-305.) It deserves to be noted, however, that special and general relativity arose primarily as a result of Einstein working on his own, in marked contrast to quantum theory, which arose much more in response to empirical considerations, and was the work of many physicists.

6. For an indication of what the problem of induction is, see note 16 of chapter one. The problem of simplicity arises because non-empirical considerations, having to do with the simplicity, unity, explanatory power, symmetry, elegance, beauty, conceptual coherence, or "inner perfection" of theories play a role in deciding what theories are accepted and rejected, in addition to considerations of empirical success and failure. The problem of simplicity is twofold. First, it involves specifying what simplicity, or unity, or explanatory power, etc. *is*, in view, especially, of the fact that a simple (unified, etc.) theory can be turned into a complex (disunified, etc.) one, and *vice versa*, by a change of formulation.

science seems to have no impact on science itself whatsoever. All this is symptomatic of the philosophy of science being the neurotic face of science.

As far as the scientific sterility of the discipline is concerned, this is something that scientists themselves occasionally comment on, as I have already mentioned. Thus Steven Weinberg recently declared: "From time to time ... I have tried to read current work on the philosophy of science. Some of it I found to be written in a jargon so impenetrable that I can only think that it aimed at impressing those who confound obscurity with profundity. ... [O]nly rarely did it seem to me to have anything to do with the work of science as I knew it. ... I am not alone in this; I know of *no one* who has participated actively in the advance of physics in the post-war period whose research has been significantly helped by the work of philosophers" (Weinberg, 1993, pp. 133-134). And John Ziman, another theoretical physicist, some years ago commented: "the Philosophy of Science ... [is] arid and repulsive. To read the latest symposium volume on this topic is to be reminded of the Talmud, or of the theological disputes of Byzantium" (Ziman, 1968, p. 31).

Philosophers of science themselves tend to be quite unashamed about the scientific sterility of their subject. They hold that their discipline is a "meta-discipline". Its task is to describe and justify scientific practice, but not to contribute to science itself. This view of their subject is forced upon them by their adherence to standard empiricism.

Cease to be the neurotic face of science, repudiate standard empiricism and adopt aim-oriented empiricism instead, and all this changes. We may take a large part of the philosophy of science to be engaged in the tasks of specifying and justifying the aims and methods of science, and spelling out metaphysical assumptions implicit in scientific knowledge. Whereas standard empiricism decrees that this is the work of a "meta-discipline", aim-oriented empiricism, on the contrary, demands that this must be pursued as a vital, integral part of

---

Second, it involves *justifying* giving persistent preference to simple or unified theories to complex or disunified theories in science, even against empirical considerations. Neither problem can be solved granted standard empiricism; both problems can be solved granted aim-oriented empiricism. To this I should add that the problem of induction *is* just the two problems of simplicity. Solve these two latter problems, and the problem of induction is automatically solved as well. For solutions see the appendix.

science itself. Untestable metaphysical assertions, at levels 3 to 7 in the hierarchy of assumptions (see Fig. 1.2) are not distinct from science; they are basic items of scientific knowledge. Methodological principles, such as symmetry principles, associated with these assumptions, represented by slanting dotted lines in Fig. 1.2, governing the choice of theories in physics, are a vital part of science itself. According to aim-oriented empiricism, the activity of trying to improve the aim and methods of science (in the light of improving scientific knowledge and understanding) is a vital part of scientific work itself: the ability to improve its aim and methods in this way is a basic feature of scientific rationality, a vital part of the reason for the amazing success of science.

The transition from standard to aim-oriented empiricism, in short, transforms the philosophy of science. Its character, its relationship to science itself, its ability to contribute fruitfully to science, are all transformed. Instead of being the discipline which actively helps sustain the neurosis of science, the philosophy of science has the chance to take a leading role in liberating science from its neurosis.

And as a bonus, long-standing unsolved problems in the philosophy of science, most notably problems of induction and simplicity, and the problem of the nature of scientific method, which have not been, and cannot be, solved granted standard empiricism, can be solved within the framework of aim-oriented empiricism. As far as induction and simplicity are concerned, I have already indicated, in the last chapter, how aim-oriented empiricism solves these problems; for a more detailed discussion, see the appendix (and also Maxwell, 1998, chs. 3-6, 2002a, and 2004d). Let us now see how aim-oriented empiricism brings clarity to questions about the nature of scientific method.

So far, aim-oriented empiricism has been discussed as a view about theoretical physics. What grounds are there for holding that this view applies to the whole of natural science, to such diverse sciences as astronomy, chemistry, biology, geology, the study of animal behaviour?

To begin with, there is one way in which all these diverse branches of natural science need to be accommodated within the framework of aim-oriented empiricism, as discussed so far. Theoretical physics is the fundamental science, as far as scientific explanation is concerned.[7] All

---

7. Physics is fundamental, given the aim of improving scientific knowledge and

the other sciences presuppose physics, where relevant, but physics does not presuppose any other science. Just as physics presupposes metaphysics (according to aim-oriented empiricism), so chemistry, astronomy and geology presuppose physics, and biology presupposes chemistry. (To say this is not to say that biology can, even in principle, be reduced to physics;[8] it is to say merely that biology, where relevant, presupposes the results of physics, but not vice versa.) All phenomena studied by natural science, whatever else they may be, are also *physical* phenomena. Because of the explanatorily fundamental, and all-encompassing, character of physics, aim-oriented empiricism, as discussed so far, is relevant to the whole of natural science, and not just to physics.

But there is a second, and even more important, way in which aim-oriented empiricism applies in detail to all the diverse branches of natural science, and thus clarifies the nature of scientific method.

Standard empiricism assigns a fixed aim to the whole of science (namely, acquisition of knowledge of factual truth), and a fixed method (namely, assess potential contributions to knowledge impartially with respect to evidence). But this generates a number of problems concerning the nature of scientific method.

First, as we have seen, it does not work: considerations of simplicity, unity or explanatory power have to be appealed to, in addition to empirical considerations, in order to account for the way theories are accepted and rejected in physics. This in turn means that metaphysical theses, in addition to empirical considerations, govern acceptance and rejection of physical theories.

Second, standard empiricism is quite unable to say what simplicity, unity or explanatory power *are*; as a result, those methods of science which appeal to these things are at best vague, at worst incoherent.

Third, standard empiricism is unable to do justice to the different methods employed by different branches of the natural science. Whereas physics is primarily experimental, astronomy and cosmology are, in the nature of things, observational. Astronomy, cosmology and geology all have an important historical character to them, something

---

understanding. It is not fundamental, given the aim of enhancing the quality of human life, as we shall see in the next chapter.

8. Elsewhere I have argued that biology (and our human world too), though compatible with physics, is not reducible to physics: see (Maxwell, 2001c, especially chs. 1 and 5-7).

foreign to theoretical physics (which seeks to discover unchanging, universal laws). The different branches of natural science tackle different fundamental problems, and seek to develop different kinds of theories. Geology and astronomy aim to acquire knowledge of particular objects (the earth, the solar system, stars and galaxies), whereas physics eschews such knowledge in that it is exclusively concerned to discover universal laws. One important task of biology is to discover how features of living things contribute to their capacity to survive and reproduce (Darwin's theory of evolution assuring us that living things have been well-designed by evolution to survive and reproduce). No such theoretical problem exists within the non-biological, physical sciences. Again, different sciences employ radically different observational and experimental methods. Standard empiricism cannot begin to do justice to this diversity of methods.

Finally, the methods of science vary, not only from science to science at any given time, but also within any given science with the passage of time. The methods of Aristotelian physics differ from those of Galilean physics, which differ from those of post-Newtonian physics, which differ again from post-Einsteinian and post-quantum mechanical physics. And it is not just in physics that methods, both theoretical and empirical, evolve with evolving knowledge; this happens in other sciences too, in chemistry, astronomy and biology. Some of the most fruitful contributions to a science amount to the discovery, the invention, of new instruments, new research *methods*.

Attempts to do justice to these features of scientific method within the framework of standard empiricism have not met with success. Disagreement and confusion about the nature of scientific method have been the outcome. Inductivists stress that laws and theories must be arrived at cautiously by generalizing from observed phenomena. Popper (1959, 1963) denies that this ever happens, and argues that scientists must boldly speculate and seek to refute their speculations experimentally. Kuhn (1970) stresses that most of the time scientists protect the established theory or paradigm from refutation, normal science being devoted to showing that the established theory is able successfully to predict phenomena. Lakatos (1970) depicts science as consisting of competing research programmes, competing fragments of Kuhn's normal science. Feyerabend (1978, 1987) concludes that none of the available views does justice to all good science, and that therefore the best policy is to hold that "anything goes". Some scientists and sociologists and historians of science have concluded that there is no such thing as scientific method, or have argued that social and political factors have to be invoked to make sense of science. More recently, Dupré has argued for a radical disunity, a radical diversity, of methods among the sciences, holding that "science is best seen, in Wittgenstein's valuable phrase, as a family resemblance concept" (Dupré, 1993, p. 10). Sokal and Bricmont (1998, p. 56) have recently put the matter like this: "there does not exist (at least at present) a complete codification of scientific rationality, and we seriously doubt that one could ever exist".

Standard empiricism, as I have said, really requires there to be unity of method throughout the sciences. The more that diversity of methods is recognized, from science to science, and from time to time, so the greater is the violence done to standard empiricism, and the more difficult it becomes to distinguish authentic science from pseudo-science. If a thousand methodological flowers should be encouraged to flourish, why should not those of astrology, alchemy and scientology be permitted to flourish as well?

All the above difficulties concerning the nature of scientific method disappear once aim-oriented empiricism is adopted.

To begin with, as far as the first two of the above difficulties are concerned, aim-oriented empiricism holds that new theories, in order to be acceptable, must accord, as far as possible, with *two*, equally important methodological requirements: (1) empirical success, and (2) compatibility with the most fruitful metaphysical thesis concerning the comprehensibility of nature currently available. Requirement (2) specifies what unity, or explanatory power, means (at any given time) as far as physics is concerned.

As far as the difficulty concerning diversity of method, from science to science, and from time to time, is concerned this, far from being a problem, is actually demanded by aim-oriented empiricism in the interests of methodological unity, rationality and scientific progress.

Two points are, here, of decisive importance.

First, as we have already seen, different branches of the natural sciences are not isolated from one another: they form an interconnected whole, from theoretical physics to cosmology, molecular biology, neurology and the study of animal behaviour. Each science presupposes relevant results from more fundamental sciences, just as physics presupposes metaphysics. But even though the different branches of natural science form an interconnected whole in this way, nevertheless different sciences have different specific aims, associated with different tasks, and different specific theoretical presuppositions. Different specific aims and presuppositions imply, immediately, different methods. Nevertheless, at a much greater level of generality, all these branches of natural science have a common aim, and therefore common methods: to improve knowledge and understanding of the natural world.

The second decisive point is this. Aim-oriented empiricism, in requiring there to be a hierarchy of aims and methods, presuppositions and methods, whenever aims or presuppositions are problematic, is able to do justice both to the *diversity* of specific aims and methods from science to science, and from time to time, low down in the hierarchy of aims and methods, and to *common* aims and methods high up in the hierarchy. Any given branch of natural science — whether it be molecular biology, geology, or the study of animal behaviour — has methods distinctive of that particular science at that particular time because of the specific presuppositions and aims of that science (at that time); these are low down in the hierarchy of aim-oriented empiricism for the science in question. But higher up in the hierarchy, presuppositions, aims and associated methods are common to all branches of natural science, in part because of the common presuppositions of physics, in part because, at a high enough level, there is the common aim of seeking knowledge and understanding of nature.

Thus all the diverse branches of natural science put aim-oriented empiricism into practice (with theses at the various levels depicted in Fig. 1.2 appropriately reinterpreted): here, there is methodological unity. But different branches of natural science, even different branches of a single science such as physics, chemistry or biology, have, at some level of specificity, different presuppositions and aims, and hence different methods. All put the hierarchical structure of aim-oriented empiricism into practice, but because different scientific specialities have different specific presuppositions and aims, at the lower end of the hierarchy, different specialities have somewhat different methods, even though some more general methods are common to all the sciences. All natural sciences apart from theoretical physics presuppose and use results from other scientific specialities (as when chemistry presupposes atomic theory and quantum theory, or biology presupposes chemistry). As we have seen, the results of one science become a part of the presuppositions of another.[9] This further enhances unity throughout diversity, and at the same time helps explain the need for diversity of method. It also helps account for the evolution of the methods of all the sciences in time. As scientific knowledge improves, so presuppositions of some speciality, taken from some more fundamental science, will tend to improve; and this in turn will lead to the improvement of the methods of that science. As far as physics is concerned, as our knowledge and understanding of the universe improve, so metaphysical ideas about how the universe is comprehensible improve as well, and this in turn leads to improvement of methods. This interplay between improving knowledge, and improving aims and methods, improving knowledge–about–how–to–improve–knowledge, manifest throughout all the diverse branches of natural science is, according to aim-oriented empiricism, a vital feature of scientific method and rationality which helps account for the explosive growth of modern scientific knowledge.[10]

In short, in order to exhibit the rationality of the diversity of method in natural science, apparent in the evolution of methods of a single science, and apparent as one moves, at a given time, from one scientific speciality to another, it is essential to adopt the meta-methodological, hierarchical standpoint of aim-oriented empiricism, which alone enables one to depict methodological unity (high up in the hierarchy) throughout methodological diversity (low down in the hierarchy). Standard empiricism, lacking this hierarchical structure, cannot begin to do justice

---

9. The presuppositions of a science, taken (in part) from more fundamental sciences (and ultimately from physics), determine what it is for a theory to be simple, unified, or explanatory within that science. These presuppositions determine the quasi non-empirical methods of that science, much as methodological principles of physics, associated with metaphysical theses at levels 3 to 7 (and represented by slanting dotted lines) of Fig. 1.2, constitute the quasi non-empirical methods for theoretical physics. (See Maxwell, 1998, 110-113, for related, additional comments.)

10. One of the great insights of Thomas Kuhn was that scientific *methods* and scientific *paradigms* are tied to each other, so that when a scientific revolution comes about and a new paradigm is adopted, the methods of the science in question change as well: see (Kuhn, 1970, chs. IV, V, and IX). But Kuhn then spoils this insight by insisting that there is no progress in knowledge across revolutions. As a result, he cannot stress the point, emphasized by aim-oriented empiricism, that methods *improve* with improving knowledge.

to this key feature of scientific method, unity throughout diversity; nor can it begin to do justice to the rational *need* for this feature of scientific method.

2.3 From Physics to Natural Philosophy

One upshot of abandoning standard empiricism and putting aim-oriented empiricism into scientific practice instead would be that physics would become something much more like natural philosophy. When modern science began, with the work of Galileo, Kepler, Descartes, Huygens, Boyle, Newton and others, the new experimental approach to improving knowledge and understanding of nature was understood to be a development of philosophy, and in England it was called "natural philosophy" or "experimental philosophy". The publications of Galileo, Kepler and the others, intermingled material that we would recognize today as "scientific" with discussion of questions of philosophy, metaphysics, epistemology, methodology, and even religion. Only much later, in the 18th and 19th centuries, did a split develop between "science" on the one hand, and "philosophy", "metaphysics", "epistemology" and "methodology", as the neurosis of natural science became established and institutionalized.[11] Curing physics of this neurosis would heal the current gulf between physics on the one hand, and philosophy on the other: metaphysics, epistemology and methodology would again become an integral part of the quest to improve knowledge and understanding of nature. Aim-oriented empiricist physics is a dramatically improved kind of natural philosophy.

The difference between "science" and "natural philosophy" may

---

11. Some physicists have continued to take metaphysical, epistemological, and methodological issues seriously, throughout the twentieth century up to the present. Discussion of such issues comes to the fore especially in times of crisis, in connection with scientific revolutions. Thus, it is to be found in connection with the development both of Einstein's general theory of relativity, and with quantum theory. Einstein, Bohr, Heisenberg, Schrödinger, Born and many others all wrote about such matters. The crucial point, however, is that *scientific* contributions are severely distinguished and demarcated from contributions to *philosophy* or *metaphysics*, however closely related these may be to scientific issues. Standard empiricism does not forbid scientists to discuss philosophy or metaphysics: it merely insists that science should not be contaminated by such discussion.

seem somewhat esoteric, little more than terminological. There is, however, one profound difference that has far-reaching implications for our whole culture and social world. Granted standard empiricism, and the split between physics and metaphysics that it insists on, the most important theoretical ideas that emerge from physics are the testable *theories* of the discipline: Newtonian mechanics, Einstein's theories of special and general relativity, quantum theory, and so on. But granted aim-oriented empiricism there is the immensely important additional item of theoretical scientific knowledge that the universe is physically comprehensible (i.e. that it has a unified dynamic structure). This thesis is untestable, metaphysical, and thus not a part of scientific knowledge, according to standard empiricism; but according to aim-oriented empiricism, it is a more secure part of theoretical knowledge than any accepted physical theory. It is, perhaps, *the* most important theoretical idea that science (or aim-oriented empiricist natural philosophy) has come up with. It is the single most important discovery that science has made about the nature of the universe. The discovery has dramatic implications for the whole of life, for our whole understanding of ourselves, the nature of consciousness, free will, the meaning and value of human life. All these things are called into question by the discovery that the universe is physically comprehensible. A basic intellectual duty of science, properly conducted, ought to be to make clear what has been discovered, and what there is of value, associated with human life, that can be reconciled with the discovery. Elsewhere, indeed, I have argued that this is our most fundamental problem of understanding (see Maxwell, 2001c, especially ch. 1; see also note 27, and associated text, of the last chapter). A society that recognized just how fundamental is the problem of what there can be that is genuinely of value in a universe presumed to be physically comprehensible might succeed in getting priorities into focus rather better than our present world, and might be less inclined to lose itself in distractions, false ideals and various religious fanaticisms and absurdities.

But granted standard empiricism, all this disappears from view. The profoundly important scientific discovery that the universe is physically comprehensible ceases entirely to be scientific, and becomes no more than a philosophical speculation.[12]

---

12. This book (like much of my writings) is a plea for greater intellectual honesty. Not everything that intellectual honesty reveals is immediately comfortable or consoling

Another gain to be had from putting aim-oriented empiricism explicitly into scientific practice is that much more emphasis would be given to the search for explanation and understanding. According to aim-oriented empiricism, that a new theory should be unified or explanatory is as important as that it should be empirically successful: both are necessary for acceptance. Standard empiricism, by contrast, places far greater stress on the importance of empirical success, and is unable to say what unity, or explanatory power, *is* (because to do so would involve committing physics to the metaphysical thesis that there is unity in nature, which violates the fundamental tenet of standard empiricism). The good sense of physicists has tended to prevail, in scientific practice, over the inadequacies of standard empiricism, and as a result most of the great theories of physics are marvels of unity and explanatory power. This is true of Newtonian theory, James Clerk Maxwell's theory of the classical electromagnetic field, and Einstein's theory of general relativity. But there are exceptions, the most dramatic being quantum theory. As I have said elsewhere, this is the best of theories, and the worst of theories.[13] The best, because of its astonishing predictive power. No other physical theory predicts such a wide range of diverse phenomena with such accuracy. The worst, because the theory, given its orthodox or Copenhagen interpretation, is seriously lacking in unity and explanatory power. Because of its failure to specify what quantum entities, such as electrons, photons and atoms, really are, in view of their apparently contradictory wave-like and particle-like features, orthodox quantum theory is formulated as a theory about the results of performing *measurements* on such quantum entities. This means, as Niels Bohr always stressed, that orthodox quantum theory must employ some part of classical physics for a treatment of measurement, as a matter of principle. Thus quantum theory consists of two incompatible parts stuck artificially together: quantum postulates, and some part of classical physics, for measurement. This theory is, as a result, very seriously lacking in unity and explanatory power. As I have argued at length elsewhere, it is imprecise, ambiguous, *ad hoc*, non-explanatory,

---

although, in the end, as I hope will emerge during the course of this book, much of real value, actual and potential, will emerge.

13. See (Maxwell, 1993b, p. 362).

incapable of being unified with general relativity, and inapplicable to cosmology.[14] In short, orthodox quantum theory, despite its very great empirical virtues, is seriously lacking in non-empirical virtues of unity and explanatory power, just the virtues emphasized by aim-oriented empiricism and fumbled by standard empiricism.

Many physicists today now acknowledge these points. But for decades after the birth of orthodox quantum theory in the 1920s, these inadequacies of the theory were vehemently denied by the majority of physicists.[15] Standard empiricism, being widely accepted, encouraged such denial. If aim-oriented empiricism had constituted orthodoxy during this period, such denial would have been impossible. General awareness of the serious theoretical and explanatory defects of orthodox quantum theory would have encouraged physicists to try to develop and experimentally corroborate a more adequate version of the theory, something which we might now have had, but which, as it is, we do not have. The theoretical defects of orthodox quantum theory which we have inherited from the 1920s are still manifest in the very latest versions of quantum theory — in quantum field theory and even in M-theory. Here, then, is one dramatic way in which even hypocritical allegiance to standard empiricism by physicists has damaged physics itself.

Yet another important (and related) difference between standard empiricist science and aim-oriented empiricist natural philosophy has to do with education. Granted the former, science education is, quite properly, severely dissociated from such things as philosophy and metaphysics. But granted the latter, philosophical problems associated with the metaphysics of science become a vital part of any decent science education. Doing science at school and university would cease to be merely the acquisition of expert knowledge and skills, and would

---

14. This is an argument I have developed over many years, at the same time developing an alternative to the orthodox version of quantum theory, free of its defects, and with slightly different, but as yet untested, predictive consequences: see (Maxwell, 1972, 1973a, 1973b, 1975, 1976b, 1982, 1988, 1993b, 1993c, 1994b, 1998, ch. 6, 2004c). A similar argument, critical of orthodox quantum theory, was developed independently by John Bell: see (Bell, 1987).

15. But it must be noted that Einstein, de Broglie and Schrödinger, who made major contributions to the creation of quantum theory, objected strongly to the orthodox interpretation. Other contemporary physicists who objected include von Laue, Bohm and Landé.

become, in addition, the exploration of some of the most exciting and dramatic problems confronting our understanding of our world and ourselves. Scientists sometimes bemoan the fact that children are not more interested in pursuing science, and poor science education is often blamed. It does not occur to scientists that their own view of science may be in part to blame.

Doing natural philosophy rather than science at school and university would even make it possible to organize much of education around the central and fundamental problem, discussed briefly in chapter one: how can we understand our human world, embedded in the physical universe, in such a way that justice is done both to the richness, meaning, and value of human life on the one hand, and to what modern science tells us about the physical universe on the other hand? It would be important to treat the problem as open and unsolved (no one knows the answer although many different wild ideas have been proposed). Education could involve encouraging students themselves to explore aspects of this problem, so that when they come to learn of the ideas and contributions of Newton, Descartes, Darwin, Faraday, Einstein and many others, these figures would be welcomed as co-workers from whom one can learn how to improve one's own ideas. The central problem could lead on both to the humanities, the arts, and social inquiry, on the one hand, and to cosmology, biology, palaeontology, physics, neuroscience, and anthropology, on the other. Natural philosophy might eventually become a central part of our education and culture.

2.4 Science and Values

So far we have considered two rival possible aims for science: the (neurotic) aim of improving knowledge about the world, nothing being permanently presupposed about the world independent of evidence; and the (real) aim of improving knowledge about the world which is presupposed to be comprehensible (or, more accurately, about which a hierarchy of increasingly insubstantial assumptions are made, including the assumption that the universe is comprehensible).

But this latter aim of seeking *explanatory truth* (as we may call it) is a special case of the more general aim of seeking *valuable truth*. Knowledge of truth that enables us to explain and understand is of great value; but not all knowledge sought and acquired in science is of value in this kind of way. There is also knowledge of truth that enables us to *do* things of value, to realize human goals of value, most notably via technological applications. There is the multitude of applications of scientific knowledge throughout almost all aspects of modern life: health, agriculture, industry, architecture, transport, communications.

It is by no means the case, it should be noted in passing, that new technological applications of science invariably come about as spinoffs of prior theoretical discoveries, discoveries of explanatory knowledge. Technological discoveries may use only long-established theoretical knowledge; some technological discoveries are made before there is a satisfactory theoretical explanation for the phenomena or processes that the technology exploits; and, on occasion, new technology may actually stimulate the subsequent development of associated theoretical science (a famous example being the steam engine stimulating the subsequent development of thermodynamics).

Standard empiricism, the neurotic conception of science, excludes metaphysics from science, as we have seen. Even more firmly, it excludes *values* from science. Viewed from the perspective of standard empiricism, it seems that values (apart from narrowly intellectual, scientific values) could only exert a corrupting influence on science. They could only lead scientists to accept some result as true because it is deemed desirable, or reject some result as false because it is deemed undesirable. Thus the thesis that there are no statistical differences in intelligence between men and women, let us say, might be deemed to be true, independent of evidence, on the grounds that it is desirable that it should be true.

If such considerations of desirability are allowed to influence scientific decisions about truth and falsity, then science is indeed subverted. But the moment the neurotic, standard empiricist picture of science is rejected, and aim-oriented empiricism is adopted instead, it becomes obvious that there is another, entirely legitimate, indeed indispensable way in which human values influence science. Values, quite properly and inevitably, influence the *aims* that scientists pursue.[16]

---

16. Given the standard empiricist picture of science, accepted laws and theories, and empirical data, are a part of knowledge, but ideas about *aims* are not; it is thus not at all obvious how values are to influence the *content* of science, apart from illegitimately influencing acceptance and rejection of laws, theories, or empirical data. Given the aim-

The neurotic, standard empiricist idea that science should seek to improve knowledge of factual truth without judgements about what is desirable or of value influencing what truth is sought is impossible to fulfil. The number of facts out there awaiting potential investigation is infinite. The entire scientific community could devote itself to acquiring knowledge about a single matchbox, if it so chose: its composition, history, manufacture, exact history of each constituent atom, etc., etc. Inevitably, scientists must choose to investigate certain facts and phenomena, and ignore others.

This is not just inevitable; it is desirable. We want science to acquire useful or valuable knowledge. It is built into the very notion of scientific knowledge, that it is knowledge that has reached a certain threshold of significance. In order to be published, it is not enough that a scientific paper establishes a new result; in addition the result must be deemed to be sufficiently important for the paper to be judged worthy of publication. A science which amassed knowledge of irredeemable trivia, would not be judged to be making splendid progress; it would be judged to be stagnating. Values are thus built into the very notions of scientific knowledge and scientific progress.

A proper, basic aim of science, then, is to improve knowledge of *valuable truth* — the aim of improving knowledge of *explanatory truth* being a special case of this.

The orthodox, neurotic, standard empiricist perspective may encourage the view that human values legitimately influence technological or applied science, but exercise no legitimate influence over pure science. Again, this is nonsense. Knowledge sought for its own sake is sought because of its human interest or significance. This is true of explanatory knowledge, of great theories of science that help us to explain and understand broad features of our world; but it is also true of, for example, knowledge sought because of its particular relevance to human life, such as knowledge about human origins and development, or the origins of life. Science is not interested, uniformly, in the contents

---

oriented empiricist conception of science, on the other hand, ideas about *aims* are a part of the content of science. It becomes obvious that values may influence aims, but not acceptance or rejection of laws, theories or evidence. It must be admitted, however, that even in the context of aims, values legitimately influence research aims, but not judgements about what is true and false.

of every chunk of spacetime throughout the history of the universe: it is especially interested in highly significant or unusual chunks, much less interested in other chunks. Counting grains of gravel on paths, or leaves on trees, is of no interest in itself whatsoever, even though this might add to the store of human knowledge.

Here, then, is a second, and in some respects much more serious neurosis of science: repression of the real aim of seeking *valuable truth* and its replacement by the officially recognized, neurotic, false aim of seeking *truth as such*, devoid of considerations of human value.

The aim of seeking valuable truth is, if anything, even more problematic than the aim of seeking explanatory truth. What is of value? Of value to whom? Who is to decide? How can conflicting values, conflicting needs, be resolved? How can we know what there is to be discovered, that science is capable of discovering, that is of value? What will it be important for us to know in 50 years time, or 150 years time?

In order to free science of its neurotic repression of values, the first step that needs to be taken is to make explicit, within the intellectual domain of science (that is, within scientific journals, texts, conferences, undergraduate and graduate courses, lectures and seminars), *both* what it is conjectured is scientifically discoverable *and* what it is conjectured would be of human value to discover: see Fig. 2.1. The idea, here, is that, as a result of making explicit conjectures concerning these two highly problematic domains, it will become easier to make a good choice of that even more problematic region of overlap between these two domains: that which is *both* scientifically discoverable *and* of value to discover: see Fig. 2.1. Precisely because this region of overlap is so highly problematic to discover (in that it involves making guesses both about what is scientifically discoverable and of value), we need to create, as an important part of scientific research, a tradition of proposing and critically assessing ideas for future research aims, plus the critical assessment of existing research aims. There need to be scientific journals devoted to the attempt to improve the research aims of science.

Scientists may be in the best position to make good judgements about what there is that is scientifically discoverable; they are not, however, necessarily in the best position to decide, for the rest of humanity, what is of value. It is above all here, in connection with values influencing aims of research, that non-scientists must contribute to science itself. Science ceases to be objective if this does not happen.

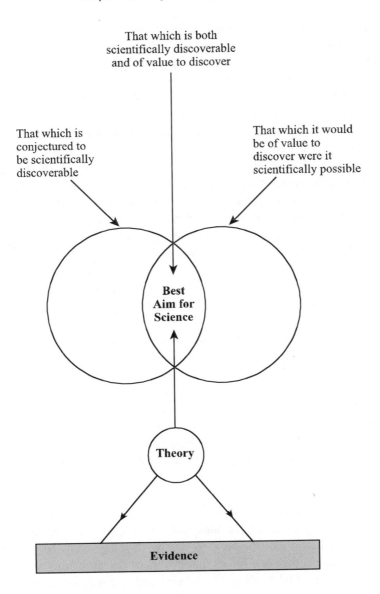

Figure 2.1: Science Seeking Valuable Truth

To say this is not to say, of course, that questions of what is of value can be decided democratically, by a vote, a poll, by market research, or by a committee of experts. Here, as in other parts of science, ideas, proposals, arguments, criticisms must be subjected to a good process of open critical discussion and assessment, so that it is the *best* ideas that come to be adopted by the scientific community.

But how can the "best" ideas be decided upon in the realm of value, that does not just prejudge the issue — the "best" ideas about what is of value corresponding, simply, to the values of those who are in charge of deciding what is and is not published, what is and what is not considered and adopted? How, to put the question slightly differently, can value-questions be decided objectively and rationally in a society which fails dismally to do any such thing — it being even uncertain as to what such decision-making about questions concerning what is of value would *mean*?

The answer is to do for questions of *value* what I have already argued needs to be done for questions of *metaphysics*. We need to create a hierarchy of conjectures as to what is of value, these value-conjectures becoming increasingly insubstantial, increasingly unproblematic and uncontroversial as we ascend the hierarchy. In this way, we can create a framework within which the rationally cooperative discovery of what is of value can become possible. It becomes possible for humanity to *learn* what is of value — as we shall see in the next chapter.

Those inclined to see science in terms of standard empiricism will, of course, deplore the suggestion that values should be incorporated into science, on the grounds that this can only undermine the objectivity and rationality of science.[17] The true state of affairs is actually all the other way round. Those who seek to exclude value-questions from science undermine the objectivity and rationality of science. As I have emphasized above, values are inevitably a part of science, simply in influencing what scientists seek to try to develop knowledge about. What is at issue is this: are values that influence research aims repressed, so that they cannot be explicitly examined and, we may hope, improved

---

17. For a typical standard empiricist critique of the idea that values should influence science see (O'Hear 1989, pp. 223-232). O'Hear takes it for granted that, in influencing science, values can only influence judgements concerning truth and falsity, which is clearly illegitimate. He fails to appreciate that it is inevitable and desirable that values should influence research aims, what scientists decide to try to develop knowledge about.

(which subverts objectivity and rationality)? Or are such values acknowledged, there being an attempt to put forward and critically assess ideas about what values should influence research aims, in an attempt to improve values and aims? The choice is between repression, neurosis, dogmatism, and the prohibition of rational discussion on the one hand, and open acknowledgement of values influencing research aims, and the sustained attempt to improve such values by means of reason, that is by means of conjecture and criticism, on the other hand. The latter option enhances the objectivity and rationality of science — the capacity of science to develop knowledge that really is of human value.[18]

At this point it may be objected that it is widely accepted that values should influence scientific research. There are, these days, committees which assess new research projects for their ethical implications, especially in the fields of animal experimentation and genetic research. Some lines of research are explicitly forbidden, because they are judged immoral. Does not this mean that the scientific community, and others, fully recognize that values play an important role in influencing choice of research aims? This would seem to be further borne out by the simple point that scientific research is expensive; scientists have to apply for grants for the research they wish to do, and are well aware that value considerations, of one kind or another, influence decisions of funding bodies. Drug companies, and other companies engaged in scientific research, support that research which will, it is judged, lead to profit for the company. Scientific funding bodies, charged with the task of distributing government money to research conducted in universities and research institutes, must inevitably take questions of value into account in deciding what kinds of research should receive support, and what kinds should not. All this, it may be argued, shows that it is common knowledge that values influence research aims. In so far as standard empiricism denies that values influence research aims, it cannot be a view that is accepted by the scientific community.

My reply to this objection comes in two stages.

First, as I pointed out in note 4 to chapter one, it is important to

---

18. Two recent works by philosophers of science that discuss the role of values in science are (Lacey, 1999) and (Kitcher, 2001). Both authors had read (Maxwell, 1984a).

recognize that standard empiricism distinguishes between the "context of discovery" and "the context of justification". According to standard empiricism, it is only in the context of justification, in the context of deciding which potential contributions to science should be accepted as constituting scientific knowledge, that evidence must be the crucial deciding factor, and all appeal to such things as metaphysics and values must be ruthlessly excluded. In the context of discovery, by contrast, there are no rational methods. Here, all sorts of considerations may legitimately influence choice of research aims: metaphysical ideas, values, dreams, ambition, competitiveness, career hopes, commercial interests and political priorities. That ethical committees, commercial interests and political priorities influence research aims, in the context of discovery, is thus fully in accord with standard empiricism.

My second point is that discussion of values plays a far more influential and radical role in science, granted aim-oriented empiricism, than it does, granted standard empiricism. According to standard empiricism, the aim of science in the context of justification is truth as such; values have no role to play here, and scientific rationality is designed to attend only to questions of fact, evidence and logic, all questions of value being excluded. But according to aim-oriented empiricism, the aim of science, even in the context of justification, is *valuable truth* (explanatory truth being a special case of this). Values have a crucial, rational role to play in deciding what is to be accepted as constituting scientific knowledge. As I have made clear above, values, of course, play no role in deciding what is true and false. They do however play the crucial role in deciding whether a new item of knowledge is sufficiently *important* to be deemed worthy of becoming a part of scientific knowledge. Values have a *rational* role to play in science. Discussion of value questions becomes an integral part of the intellectual content of science. Such discussion needs to take place in scientific journals, textbooks, seminars, lectures, and science courses in universities and schools.

The two views, in short, give very different roles to the discussion of values in science. Just how different can be gleaned from the following summary.

Standard Empiricism

(i) Values influence science only in the non-rational context of discovery.

(ii) The aim of science, in the context of justification, is truth as such.

(iii) Values have no role to play in the context of justification.

(iv) They have no rational role to play in science.

(v) Discussion of questions of value has no place within the intellectual content of science — scientific journals, texts, papers, lectures, etc.

(vi) Scientific contributions and discussion takes place, essentially, at just two levels, namely at the levels of (1) evidence, and (2) theory, values having no role to play at either level.

(vii) There can be no such thing as the sustained attempt to *improve* the overall aims

Aim-Oriented Empiricism

(i) Values legitimately influence science in *both* contexts of justification and discovery.

(ii) The aim of science, in the context of justification, is *valuable* truth, with *explanatory* truth as an important special case.

(iii) Values have an important role to play in the context of justification.

(iv) They have a rational role to play in science.

(v) Discussion of questions of value has an important place within the intellectual content of science — scientific journals, texts, papers, lectures, etc.

(vi) Scientific contributions and and discussion takes place at *three* levels (at least), namely at the levels of (1) evidence, (2) theory, and (3) aims, metaphysical ideas and values having an especially important role to play at the third level of aims. (More levels are required for the task of improving aims.)

(vii) It is important to create and sustain the attempt to *improve* the overall aims and priorities of

and priorities of science, from the standpoint of value, in a rational manner.	science, from the standpoint of value, in a rational manner, by proposing and critically assessing possible and actual aims for science.
(viii) No purpose would be served by having decisions about funding of scientific research subjected to sustained critical scrutiny in journals devoted to that task, and in the media.	(viii) An important purpose would be served by having decisions about funding of scientific research subjected to sustained critical scrutiny in journals devoted to that task, and in the media.
(ix) Non-scientists can contribute to science only in the context of discovery, in contributing to decisions about the morality of some research proposals.	(ix) Non-scientists can contribute to science in both contexts of discovery and justification, by contributing to discussion of aims, at level (3).

Standard and aim-oriented empiricism disagree on all nine points. The really crucial difference (to which the other differences contribute) is point (vii) concerning the possibility of improving, by rational means, the aims of science, in the context of justification, from the standpoint of their value to humanity. This simply does not arise, given standard empiricism. Given aim-oriented empiricism, it becomes an important, if undoubtedly immensely difficult, task and obligation for the scientific community.

Values are problematic quite generally. There are, however, problems about values associated with science that are quite specific to science, due to the fact that conflicting interests and values will, almost inevitably, be built into the scientific enterprise.

There is, on the one hand, the noble idea that science should be for humanity as a whole, and should not merely serve the interests of the wealthy and powerful. One could regard this as meaning that science should give priority to those whose needs are greatest, the poor, the sick, the deprived, the enslaved and exploited. Such an idea goes back to the creation of modern science. As I put it some year ago: "Robert Boyle, one of the founding fathers of modern science, had this to say about what he called the 'Invisible College' — a sort of forerunner of the Royal Society, and thus of organized scientific research. 'The 'Invisible College' [consists of] persons that endeavour to put narrow-mindedness out of countenance by the practice of so extensive charity that it reaches unto everything called man, and nothing less than an universal good-will can content it. And indeed they are so apprehensive of the want of good employment that they take the whole body of mankind for their care.'" (Maxwell, 1984a, p. 54).

However concerned scientists may be that priorities of research should reflect the kind of priorities that Boyle here expresses — the priorities of the poor and the sick, the priorities of the great body of humanity — it is all too likely that in reality research will reflect the interests of those who fund research: industry and government. It is by no means obvious that the interests of industry and government in industrially advanced, wealthy parts of the world coincide with the interests of that large majority of humanity living in the poor, underdeveloped parts of the world.

On top of this there are the interests of scientists themselves, who may be more concerned to pursue research projects that advance understanding rather than alleviate the suffering of the poor of the third world. Scientists may have noble aspirations in pursuing pure research; and they may have less noble ones, that have to do with achieving results of high scientific prestige which win scientific prizes, or at least advance scientific careers.

Only open, honest, critical exploration of such conflicts of interests and values is likely to lead to a science that genuinely serves the best interests of humanity. Standard empiricism, repressing such discussion, is likely to lead to research that serves the interests of scientists themselves, and the interests of the wealthy and powerful, under a cloak of rhetorical rationalization about science being wholly objective, factual and value-neutral, and in this way truly serving the interests of humanity.

The priorities of medical research illustrate the points just made. Most medical research is conducted in wealthy countries; much of it is conducted by, and funded by drug companies. Not surprisingly, the priorities of the research tend to respond to health problems of people living in wealthy countries, and to the interests of drug companies (profit). Diseases that afflict poor people living in poor parts of the world, in Africa, Asia and South America, do not get anything like the attention they deserve.

One small step that could be taken to help build into the institutions of science sustained imaginative and critical exploration of actual and possible values and priorities of research would be to create a body that might be called the "Science and Human Need Commission", to be modelled on the Human Genetics Commission (a body that does exist in the UK). The task of the Science and Human Need Commission would be to explore the mismatch between the priorities of research on the one hand, and the priorities of human need on the other, world-wide. Its task would be to invite, develop, assess and publicize proposals for research, and research priorities, that might help diminish the mismatch. Its members would consist of scientists and non-scientists. It would need to be set up and funded by government, but its powers, initially at least, would only be advisory.

Viewed from the perspective of aim-oriented empiricism, and taking the view of Robert Boyle that a basic aim of science ought to be to serve the best interests of humanity, it is obvious that science cannot hope to be rational, and cannot hope to serve the interests of humanity properly, without possessing, as an important part of its institutional structure, something like the Science and Human Need Commission. Scientific Research Councils that make decisions about how government money is to be spent on research, cannot be expected adequately to perform the tasks of the Science and Human Need Commission. What is required is a body which can take a broader view of science and human needs and problems, not tied to decision-making, which can look at science in a global, and not just national, context, and which can review critically funding decisions made by the Research Councils, and by Companies and Charities funding research.

The fact that nothing like the Science and Human Need Commission exists at present — and the fact, even more significant, that the *need* for such a Commission is not appreciated — very strikingly demonstrates what a hold standard empiricism has over the scientific community and those who think about, and agitate in connection with, science.[19]

---

19. I have, however, suggested to Scientists for Global Responsibility (a body that takes Robert Boyle's vision to heart) that it should campaign for the creation of something like the Science and Human Need Commission. This campaign may one day succeed.

2.5 Science and Politics

Why does science seek to improve knowledge of valuable truth? Science does this in the hope that this knowledge will be *used* by people, in their lives, to enrich the quality of their lives. There is little point in a scientist discovering something of great potential value if no one makes use of it. Science comes to life, as it were, when it is used by people, in one way or another, as a part of life. Locked away in journals, or in the notebooks, computers or heads of scientists, scientific discoveries have only potential value, as far as the body of humanity is concerned.

It needs to be appreciated that this applies just as much to "pure" science as it does to "applied". Improving our knowledge and understanding of aspects of the universe and ourselves for their own sake is of value in so far as it is the knowledge and understanding of *people*, whether scientists or non-scientists, that is improved. If science one day becomes fully automated, so that only robots (without consciousness, let us assume) can do and "understand" science then, in one sense, science might be making splendid progress. But in another, much more important sense, science would have come to an end precisely because *human beings* no longer had any knowledge or understanding of the science that was being done. Instead of enhancing people's knowledge and understanding, science would fail to contribute anything intellectual or cultural to humanity. And all this is even more obvious when it comes to the *practical* or *technological* value of science. In *both* cases, what matters is the capacity of science to contribute to the enrichment of human life, either directly by means of provoking enhanced knowledge and understanding of, or curiosity about aspects of the universe and ourselves, or indirectly by means of technological or other applications which enable people to achieve goals of value in life (such as health, travel, communication, etc.).

This means that the impersonal intellectual aim of science of improving knowledge of explanatory truth or, more generally, of valuable truth, is pursued as a *means* to the end of pursuing the *human, social,* or *humanitarian* aim of contributing to the enrichment of human life. The purely *intellectual* aims of science are pursued, we may say, in order to contribute to more fundamental *political* aims, "political" here being interpreted, not narrowly as "party politics", but broadly as the activity of seeking to improve the human condition.   Science is fundamentally a part of a *political programme* to improve the quality of human life, and the intellectual aims of science are means to that more fundamental end.[20]

But if those who view science from the orthodox, standard empiricist standpoint are inclined to throw their hands up in horror at the idea that science should include metaphysics and values, they will undoubtedly be apoplectic at the suggestion that the ultimate purpose of science is *political*! For them, the idea that science is a part of some *political programme* will be the ultimate obscenity. If taken seriously, such an idea, so they will passionately maintain and believe, will spell the end of science.

The reply to this vehement denial (the vehemence typical, of course, of neurosis) is essentially the same as before. Whether scientists like it or not, science is carried on within the human world, funded primarily by industry and government, its products used by people, industry, government and other institutions in a variety of ways in order to realize a variety of human ends, from the good to the bad. Not only is this *inevitable*; it is *desirable* that science should be used by people in life. It is both inevitable and (apart from dreadful misuses of science) *desirable* that science should be a part of a political process (in the broad sense of "political"). Repression of the human, humanitarian, social or political goals and dimensions of science amounts to no more than the denial of an undoubted reality, even if a highly *problematic* reality.

All scientists recognize, of course, that the products of scientific research are used and misused by people, by industry, by governments in a variety of ways for a variety of social, economic or political ends. Those who see science in terms of standard empiricism, however, sharply dissociate the intellectual, the purely scientific aims and aspects of science from its human uses and misuses. They see science as having the intellectual aim to improve knowledge about aspects of the world around us; they do not see *science itself* as having any humanitarian, social or political aim or function. It is here that the act of repression, the neurosis, comes in. Inevitably, and quite properly, in engaging in scientific research, teaching and publishing the results of that research, in a particular socio-economic-political context, there is a social,

---

20. For earlier expositions of this argument see Maxwell (1976a, 1984a).

economic and political dimension to what the scientist does, however much this may be neurotically suppressed. And furthermore, it is quite proper that science, even at its most "pure" and esoteric, its most theoretical and far removed from the practical, is construed as making a contribution to the welfare of humanity, to the enrichment of human life. Enhancing our scientific understanding is worthwhile in so far as it leads to, is associated with, enhancing the understanding that *people* have of the world in which they live.

The argument now proceeds in exactly the same way as before, in connection with the suppression of metaphysics and values. Denying the social, humanitarian or political dimension of science, repressing these social aims of science, means that the severe *problems* associated with these dimensions, these aims, cannot be explicitly discussed within the intellectual domain of science. The result is that scientists are more or less helpless when it comes to having their work misused by industry, governments and others. Having declared the *use* and *misuse* of science to lie outside the intellectual domain of science itself, the scientific community is ill-equipped to combat misuses of scientific knowledge, and to develop the social or institutional means for the humanly valuable use of science. Repressing the political dimension of science undermines the capacity of science to be of real value to humanity.

It may be objected that scientists understand, all too well, that there is a political dimension to science. Scientists engage actively in politics. They lobby the government for more funds for science. They form pressure groups to win favourable publicity for controversial research, such as research into genetically modified food or nanotechnology. There are scientists, and scientific organizations, that understand the need for good public relations just as well as any cabinet minister or government.[21]

My reply to this objection is analogous to my reply to the similar objection made above to the claim that orthodoxy excludes values from science; as before, my reply comes in two stages.

First, as before, it is vital to take into account the distinction between

---

21. For a fascinating account of the way science in the USA became involved with and, to some extent, corrupted by, politics after the second world war and the Manhattan Project, see (Greenberg, 1971).

the aims of science in the contexts of discovery and justification. Most scientists would readily acknowledge that politics may have a role to play in the context of discovery, in the context of getting funds for, and arousing public interest in, science. But they would fiercely deny that politics can have any legitimate role in the context of justification, in the intellectual domain of science. But above I argued that it is just in this latter context of justification, within the intellectual domain of science, that science inevitably, and quite properly, has a political dimension to it, in that a proper basic aim of all of science is to contribute to human welfare, the enrichment of human life, culturally or practically. Only in extreme circumstances should scientists distort or suppress knowledge of truth for political or humanitarian reasons: an example might be suppressing knowledge of nuclear fission in Hitler's Germany. In more normal circumstances, the political dimension of science must never affect judgements of truth and falsity (except in those circumstances when being wrong would have disastrous human consequences and this provides grounds for exercising additional caution.) The political or humanitarian dimension of science ought primarily to influence decisions about what it is scientifically important to do: what kind of research to pursue, what reported scientific results to check, what to teach, what to announce or criticize in the public domain.

Second, granted aim-oriented empiricism, the political dimension of science plays a far more radical role than it does granted standard empiricism. Not only does aim-oriented empiricism insist that *all* of science has a political aspect; it also insists that this political or humanitarian aspect should be subjected, as far as possible, to the same high intellectual standards of critical scrutiny as the rest of science. A primary task must be to try to *improve* political decision-making as this affects science and its use, culturally and practically, in social life. The two columns, above, indicating the different roles that standard and aim-oriented empiricism give to *values* in science, could be drawn up again to indicate the different roles that the two views give to the *political* dimension of science. For aim-oriented empiricism, *all* of science has, as a basic aim, to contribute to the enrichment of human life by intellectual means; it is vital that it does this with full intellectual integrity, in a rational way. Far from politics corrupting science, science needs to contribute to the decorruption of politics.

But at this point, the line of argument we have been pursuing may seem to become somewhat implausible. Is it really to be expected that scientists, in their professional capacity as scientists, will enter the political domain, do battle with the might of industry, government and public opinion in an attempt to improve the rationality of political decision-making? Would not such a task leave no time for the scientific research itself? Science might come to an end through the sheer exhaustion of its practitioners, through compassion fatigue! Or, put another way, is it not the job of *social* scientists, rather than natural scientists (and so far we have been concerned exclusively with natural science) to come to grips with the social world?

This brings us to an even more serious dimension of neurosis associated with science, this time bringing in the social sciences. It will turn out that this neurosis affects not only the academically respectable social sciences, such as economics, political science, psychology and anthropology, but also the less academically secure social science with which we began, namely psychoanalytic theory! The argument concludes by coming full circle: psychoanalytic theory itself is deeply neurotic.

Chapter Three

Implications for Social Inquiry

3.1 The Crisis of Science without Wisdom

In assessing the success of science, and whether its neurosis has adversely affected this success, much may depend on what we take the aim of science to be.

Granted that the aim is to improve our knowledge about the world, science must be deemed to be extraordinarily successful, despite its neurosis, as long as "our knowledge" is interpreted to mean "knowledge of experts". Add understanding to this aim, and science becomes slightly less successful, in that orthodox quantum theory, despite adding immensely to our knowledge of empirical phenomena, nevertheless fails to provide understanding, as we saw in the last chapter.[1] Change the aim again, and interpret "our knowledge" to mean, not "knowledge of experts", but "knowledge of humanity as a whole", and the success of science is converted into colossal failure. Most people alive today are pitifully bereft of any real knowledge and understanding of the scientific picture of the world and how we human beings fit into this picture. It is unreasonable to expect most people to possess expert knowledge of some aspect of science; but it is not at all unreasonable to suppose that most people might have some understanding of the overall picture of things that science presents us with. This is a realistic aim, but one which modern science has failed miserably to attain. (This illustrates how dramatically the success or failure of science may depend on delicate adjustments to what one takes the basic *aim* of science to be.)

If the aim of science is to improve our knowledge of humanly valuable truth then, again, natural science must be deemed to be

---

1. Orthodox quantum theory fails to solve the wave/particle dilemma, and as a result is a theory about the results of performing measurements on quantum systems. The resulting theory fails to provide understanding because (a) it is an *ad hoc*, and therefore non-explanatory theory being made up of quantum and classical parts; and (b) it fails to answer two basic questions about the quantum domain, namely "What sort of entities are quantum systems?" and "Is the quantum domain fundamentally probabilistic or deterministic?". See (Maxwell, 1972; 1976b; 1982; 1988; 1993b; 1994b; 1998, ch. 7; 2004c).

astonishingly successful (ignoring the caveat about "our" this time). But there have, of course, been some awful scientific and technological discoveries: machine guns, bombs, rockets, mines, anti-personnel mines, chemical, biological and nuclear armaments. The technology of war and death. All this is knowledge and technology we could do without; but we have to balance this with all the knowledge and technology that makes the wealth, the richness of the modern world possible.

But if the aim of science is taken to be to help promote human welfare or, more radically, to help humanity learn how to build a better world or become civilized, the answer must be, surely, that science, given this humanitarian, political aim, has not been especially successful. In countless ways, life for most people in the wealthy, industrially advanced world is, today, vastly better than life in medieval Europe, let us say, or in hunting and gathering tribes ten thousand years ago. We are healthier, we live longer, we can travel the globe, enjoy the artistic and intellectual works of previous generations, we are free from hunger, and we have all sorts of comforts and amenities which would have been beyond the wildest dreams of people living half a millennium ago. All this is due to the intervention of modern science. What makes early 21st century life so utterly different from any previous age all comes, in the end, from science — or at least has been made possible by science. This is the success.

But this success has to be balanced by horrors of the past century. There are the millions killed in savage wars, the death camps, the unspeakable acts of near genocide, the brutal dictatorships, the extreme poverty of over two thirds of people alive today, deprived of safe water, sufficient food, health care, employment. There is the explosive growth in the world's population. There is the destruction of traditional ways of life, countless societies, cultures and languages being wiped out before the relentless march of the modern world. There is global warming, the thinning of the ozone layer, and other forms of pollution. There are the mass extinction of species, and the progressive destruction of tropical rain forests and other natural habitats. And, of course, as mentioned above, there is always the threat posed by modern armaments, conventional (so-called), chemical, biological and nuclear.

The decisive point to appreciate is that *all* these distinctively modern horrors can be traced back to the intervention of modern science and technology. Without science, they would not have happened.

More precisely, all these horrors are due to the fact that we have solved the big problem of progressively improving knowledge about the world (by creating modern science) without *also* solving the big problem of making social progress towards a wise, civilized world.

Solving the first big problem of learning leads to rapidly increasing scientific knowledge which leads to (or has associated with it) rapidly increasing technological know-how, which in turn brings with it an immense increase in the power to act. In the absence of global wisdom, in the absence, that is, of a solution to the second great problem of learning, the increase in the power to act may have good consequences, but will as often as not have all sorts of harmful consequences, whether intended or not.

Thus scientific and technological progress make possible the development of modern armaments, conventional, chemical, biological and nuclear which, in the absence of global wisdom and civilization, lead to wars in which millions are killed. Improved hygiene, medicine and agriculture, by-products of scientific knowledge, make possible the rapid growth of the world's population, with all its attendant problems. Technology makes industrialization possible, which leads, amongst other things, allied to population growth, to pollution, global warming, destruction of rain forests and other natural habitats, and the mass extinction of species. Modern weaponry and mass communications, especially television, aids dictatorships in holding onto power. Science and technology has also led to the immense discrepancies in wealth and power that exist in the world today, and the domination of the world by a culture that came from Europe. Modern science was discovered in Europe in the 16th/17th centuries. As a result, it was in Europe that new lethal weapons were first developed, facilitating defeat in war of people elsewhere (America, north and south, Africa, Australia, New Zealand and elsewhere). Again, as a result, it was in Europe and its dependencies that the industrial revolution happened (made possible by science and technology), which led to an astonishing increase in wealth and the creation of the modern world. Elsewhere, in the "third world", people have experienced massive population growth, but are only recently, in some places, catching up with the industrially advanced world.

All our distinctively modern horrors, to repeat, are a direct result of the fact that we have solved the first great problem of learning —

learning about the nature of the universe — but have not yet discovered how to solve the second great problem of learning —learning how to achieve global wisdom and civilization. And the vital point to appreciate is that this combination of solved and unsolved problems is *bound* to lead to a mixture of benefit and disaster. Science and technology massively increase our power to act; without global wisdom, this is bound to have both good and disastrous consequences.

Humanity is, in other words, at present, in a situation of great peril — unprecedented peril, when judged from a historical perspective. Before the advent of modern science and technology, lack of global wisdom did not matter too much; we lacked the power to wreak too much havoc on ourselves and our surroundings. Now, with modern science and technology our power is terrifying, and global wisdom and civilization have become, not a luxury but a necessity.

There are those, of course, who blame science for our troubles. But this is to miss the point. It is not science that is to blame, but ourselves for failing to learn wisdom. Instead of blaming science, we should seek to *learn* from science — learn, in particular, from the extraordinary success of science in solving the first big problem of learning, its success, that is, in improving our knowledge and understanding of this mysterious world we find ourselves in.

Can we, in other words, learn from *scientific progress* towards greater knowledge how to achieve *social progress* towards a better, wiser, more civilized world? Can we generalize the progress-achieving methods of science, and then apply these generalized methods to the immense task of creating global civilization?

3.2 The Traditional Enlightenment

This is an old idea. It goes back to the Enlightenment of the 18th century. Indeed, this was the basic idea of the *philosophes* of the Enlightenment — Voltaire, Diderot, Condorcet *et al.*: to learn from scientific progress how to achieve social progress towards world enlightenment.

The best of the *philosophes* did what they could to put this immensely important idea into practice, in their lives. They fought dictatorial power, superstition, and injustice with weapons no more lethal than those of argument and wit. They gave their support to the virtues of tolerance, openness to doubt, readiness to learn from criticism and from experience. Courageously and energetically they laboured to promote rationality in personal and social life: see (Gay, 1973).

Unfortunately, in developing the Enlightenment idea intellectually, the *philosophes* blundered. They developed the Enlightenment programme in a seriously defective form, in a *neurotic* form, and it is this immensely influential, defective, neurotic version of the programme, inherited from the 18th century, which may be called *The Traditional Enlightenment*, that is built today into our institutions of inquiry. Our current traditions and institutions of learning, when judged from the standpoint of helping us learn how to become more enlightened, are defective, neurotic and irrational in a wholesale and structural way, and it is this which, in the long term, sabotages our efforts to create a more civilized world, and prevents us from avoiding the kind of horrors we have been exposed to in recent times — wars, third-world poverty, environmental degradation. Rationalistic neurosis has become profoundly damaging!

The *philosophes* of the 18th century assumed, understandably enough perhaps, that the proper way to implement the Enlightenment programme was to develop social science alongside natural science. Francis Bacon had already stressed the importance of improving knowledge of the natural world in order to achieve social progress.[2] The *philosophes* generalized this, holding that it is just as important to improve knowledge of the social world. Thus the *philosophes* set about creating the social sciences: history, anthropology, political economy, psychology, sociology.

This had an immense impact. Throughout the 19th century the diverse social sciences were developed, often by non-academics, in accordance with the Enlightenment idea. Saint-Simon, Comte, Mill, Marx, Durkheim, Weber all contributed to this development. Gradually, universities took notice of these developments until, by the mid-20th century, all the diverse branches of the social sciences, as conceived of by the Enlightenment, were built into the institutional structure of universities as recognized academic disciplines.[3]

---

2. For the importance of Francis Bacon for the Enlightenment see (Gay, 1973, vol. 1, pp. 11-12 and p. 322).

3. For an excellent brief account of the origins of social science along these lines see

As I indicated in the Preface, *The Traditional Enlightenment* has long been opposed by the Romantic movement, and by what Isaiah Berlin has called "The Counter-Enlightenment" (Berlin, 1979, ch. 1). Whereas *The Traditional Enlightenment* valued science and reason as tools for the liberation of humanity, Romanticism found science and reason oppressive and destructive, and instead valued art, imagination, inspiration, individual genius, emotional and motivational honesty rather than careful attention to objective fact. Romanticism and the Counter-Enlightenment also had a considerable impact on the way academia developed during the 19th and 20th centuries, especially in such fields as social anthropology, social psychology, sociology, cultural studies, philosophy and history of science. Phenomenology, existentialism, postmodernism, poststructuralism, social constructivist history of science, and so-called continental philosophy can all be regarded as manifestations of the Romantic opposition and the Counter-Enlightenment: for references see the Preface.

One might almost say that academia as it exists today is the outcome of two influences: *The Traditional Enlightenment*, and the Romantic opposition to the Enlightenment — the Romantic opposition to science and reason.

But this opposition entirely misses the point. For what is objectionable about *The Traditional Enlightenment* is not its espousal of reason, but its *lack* of reason. In putting *The Traditional Enlightenment* into practice, by creating and pursuing social science alongside natural science, and restricting itself primarily to the pursuit of knowledge, academic inquiry becomes irrational to the extent that it violates *three* of the four most elementary rules of reason one can conceive of — when judged from the standpoint of helping to promote human welfare by intellectual means (as we shall see below). Furthermore, if the basic Enlightenment idea had been properly implemented by the *philosophes*, the outcome would have been a kind of *synthesis* of Rationalism and Romanticism (again, as we shall see below).

In order to implement properly the basic Enlightenment idea of learning from scientific progress how to achieve social progress towards a civilized world, it is essential to get the following three things right.

(Fargaus, 1993, Introduction). See also (Hayek, 1979).

(i)   The progress-achieving methods of science need to be correctly identified.

(ii) These methods need to be correctly generalized so that they become fruitfully applicable to any human endeavour, whatever the aims may be, and not just applicable to the endeavour of improving knowledge.

(iii) The correctly generalized progress-achieving methods then need to be exploited correctly in the great human endeavour of trying to make social progress towards an enlightened, civilized world.

Unfortunately, the *philosophes* of the Enlightenment got all three points disastrously wrong. They failed to capture correctly the progress-achieving methods of natural science; they failed to generalize these methods properly; and, most disastrously of all, they failed to apply them properly so that humanity might learn how to become civilized by rational means. That the *philosophes* made these blunders in the 18th century is forgivable; what is unforgivable is that these blunders still remain unrecognized and uncorrected today, over two centuries later. Instead of correcting the blunders, we have allowed our institutions of learning to be shaped by them as they have developed throughout the 19th and 20th centuries, so that now the blunders are an all-pervasive feature of our world.

These are the blunders that have caused the institutional neuroses from which we suffer today — the neurosis of natural science, the much more serious neurosis of social inquiry, and the even more serious neurosis of academic inquiry as a whole. Correct these intellectual/ institutional blunders, and we cure academia of its multiple neuroses, as we shall see.

3.3 The Three Blunders

So what exactly are the three blunders of *The Traditional Enlightenment*, as embodied in academic inquiry today, and what needs to be done to put them right? Let us take the three blunders in turn.

The first blunder has already been discussed at some length. It involves accepting standard empiricism instead of aim-oriented empiricism.

But what of the *second* blunder? The task here is to generalize the progress-achieving methods of science appropriately so that they become progress-achieving methods that are, potentially, fruitfully applicable to *any* worthwhile, problematic human endeavour. The task, in other words, is to generalize scientific rationality so that it becomes rationality *per se*, able to help us to achieve what is of value whatever we may be doing.[4]

Needless to say, scientists and philosophers, having failed to specify the methods of science properly, have also failed to arrive at the proper generalization of these methods. The best attempt known to me is that made by Karl Popper.[5] I now say something about Popper's attempt to put right the second blunder of *The Traditional Enlightenment*, how it improves on earlier ideas, but is nevertheless still defective, and needs to be improved. I then go on to discuss the *third* blunder of *The Traditional Enlightenment*, by far the most serious, and what needs to be done to put it right.

According to Popper, science makes progress by putting forward wildly speculative theories about aspects of the world which, though not verifiable, are nevertheless, crucially, empirically *falsifiable*. These are then subjected to a devastating onslaught of attempted empirical refutation. When such a theory is refuted by observation or experiment,

4. Rationality, as I use the term in what follows, appeals to the idea that there is some no doubt rather ill-defined set of rules, methods or strategies which, if put into practice, give us our best chances, other things being equal, of solving our problems, realizing aims of value that are in our best interests. These rules of reason are meta-rules: they presuppose that there is much that we can already do, and specify how we can best marshal our existing capacities to solve new problems, realize problematic aims. As we shall see, the rules of reason do not specify precisely what needs to be done, in a mechanical fashion; nor do they guarantee success; they merely give us a better chance of success than if we systematically violate them. The rules of reason, in the context of science, reduce to scientific method; and the methods of science, generalized to apply, in a potentially fruitful way, to all worthwhile, problematic tasks, become the rules of reason.

5. Even though it is a magnificent contribution to the Enlightenment programme, paradoxically, Popper himself did not present his work in this light (and may not have understood it in this way). As far as *The Open Society and Its Enemies* is concerned, the Enlightenment is the *ancient Greek* Enlightenment of Pericles and Socrates, not the 18th century Enlightenment of Voltaire and Diderot. One of the few writers who appreciates fully just how important it is to set Popper's early work into the context of the 18th century Enlightenment is Malachi Hacohen: see (Hacohen, 2000).

scientists seek to discover a better speculative theory, which asserts more about the world, and which is not refuted by the observations or experiments that demolished its predecessor. Theories that survive the persistent onslaught of attempted empirical refutation become a part of scientific knowledge, but all such knowledge is permanently conjectural in character. There is no verification of theory in science. Science thus makes progress by a process of conjecture and refutation, of trial and error. That a theory has survived the devastating critical onslaught hurled at it by scientific research is the best assurance that we can have that the theory deserves to be taken seriously, as our best attempt at discovering the truth. Attempts to show that such a theory has been verified are counterproductive; they cannot succeed, and can only lull us into a false sense of security, so that we play down the need for further attempted refutation (it being redundant to try to show that a verified theory is false). The vulnerability of scientific theories to empirical falsification is crucial to the capacity of science to make progress: for it means that in science we can discover that we are wrong, and that there is an urgent need to think up a better theory. Falsification provides the spur for progress.[6]

Popper went on to argue that this conception of scientific method can be generalized to form a conception of rationality, according to which one seeks to solve problems quite generally by putting forward conjectures as to how a given problem is to be solved, these conjectures then being subjected to sustained *criticism* (criticism being a generalization of attempted empirical refutation in science).[7] Not just in science, but in the rest of life too, the best that we can do in seeking to solve those problems we need to solve in order to make progress towards goals of value is to propose and critically assess possible solutions, try out various possibilities and assess the consequences of what we do critically. In his *The Open Society and Its Enemies* Popper demonstrates that this apparently almost prosaic conception of critical rationalism, if taken seriously, has far reaching consequences for politics and political philosophy, for social problem-solving, and for the social sciences (see

---

6. See (Popper, 1959; 1963; 1972, ch. 1; 1976).

7. "[I]nter-subjective *testing* is merely a very important aspect of the more general idea of inter-subjective *criticism*, or in other words, of the idea of mutual rational control by critical discussion" (Popper, 1959, 44, n *1). See also (Popper, 1963, pp. 193-200; 1972, p. 119 & 243; 1976, pp. 115-116).

Popper, 1963, 1974). As a result of bringing about a revolution in our conception of scientific method, and of rationality more generally, Popper in effect transforms the very idea of "the rational society", so that this ceases to be something that is morally and politically abhorrent, and becomes both highly desirable, and achievable, instead.

Traditional views, prior to Popper, tend to see science as establishing secure knowledge of truth by means of evidence, and tend to see reason as establishing truth by means of deductive argument. Science and reason determine truth. In terms of such traditional conceptions, the "rational society" can only be a society determined, or at least severely constrained, by "the rules of reason". Reason becomes a kind of tyrant. Individual liberty, diversity of views and ways of life, wayward imagination, disagreement and protest would all be suppressed by the iron rule of reason and logic. Granted such verificationist, authoritarian conceptions of reason, the "rational society" can only be regarded as a kind of nightmarish totalitarian state, the very opposite of democracy and liberalism.[8]

But Popper's revolutionary ideas about science and reason change all this dramatically. First, granted Popper's falsificationist conception of scientific method, imagination plays a crucial role in science. Imagination is needed to dream up new wild speculations, subsequently to be submitted to ferocious attempts at empirical refutation. Second, plurality of conflicting theories is absolutely essential for scientific progress, not only to increase the store of theories to be submitted to attempted refutation, but in order to ensure that theories are severely tested in the first place. In order to make sense of the idea of severe testing, we need to see the experimentalist as having at least the germ of an idea for a rival theory up his sleeve (otherwise testing might degenerate into performing essentially the same experiment again and again). This means experiments are always *crucial* experiments, attempts at trying to decide between two competing theories. Theoretical pluralism is necessary for science to be genuinely empirical.[9]

---

8. For a novelist's fantasy of such a "rational" society, see Zamyatin (1972).
9. This point is especially emphasized and further developed by Feyerabend (1965). Popper too emphasizes that, in order to make sense of the idea of severe testing we need to appeal to crucial experiments: see, for example, Popper (1963, p. 112).

Both these points carry over when Popper's falsificationist conception of scientific method is generalized to form critical rationalism. Reason, quite generally, is at a loss without imagination. Imagination is required to dream up possible solutions to problems, which can then be submitted to severe criticism. Again, plurality of views is an essential ingredient of Popper's conception of reason. Criticism can only deliver a good idea as to how to solve a problem if there is a plurality of ideas to criticize in the first place. And merely in order to criticize an idea, one needs to have some kind of rival idea in mind, at least as a possibility.

Rationality, as construed by Popper, requires plurality of ideas, values, ways of life, and the freedom to imagine, to criticize authority, dogma and received opinion. It demands sustained tolerance of diversity of views and ways of life, together with the existence of traditions of criticism, so that good ideas may be selected from a pool of not-so-good ideas. Reason, as Popper emphasizes, needs to be seen in social, political and institutional terms (Popper, 1963, ch. 24; 1974, section 32). Thus, granted Popper's revolutionary conceptions of scientific method and reason, the "rational society" is not some kind of totalitarian society, but just the opposite, the "open society" — a society that tolerates doubt, diversity of views and ways of life, criticism, and sustains individual liberty, reasonableness, humanity, justice and democracy. Reason, instead of being the enemy of freedom, individuality, imagination, democracy and justice, becomes the friend of these things, indeed essential for their preservation and development. As Popper puts it in a stray remark tossed out during the course of developing the argument: "We have to learn the lesson that intellectual honesty is fundamental for everything we cherish" (Popper, 1963, vol. 2, p. 59).

In his *The Open Society and Its Enemies*, Popper depicts an epic struggle between those who have sought to help sustain and promote the open society (i.e. the rational society), and those who have opposed it. And he shows how even some of the greatest thinkers of the past have been beguiled by false ideas of science and reason into arguing for the closed society, above all Plato and Marx.

The path Popper pursues, from his conjectural, falsificationist conception of science to its generalization to form critical rationalism, and its application to some of the most urgent and profound political and social problems of our times, represents, in my view, an immensely important rediscovery and transformation of the Enlightenment programme we have inherited from the 18th century, of learning from scientific progress how to achieve social progress towards an enlightened world. Popper's contribution is important and profound; but it is nevertheless defective. It needs further improvement.

Popper's conception of scientific method is defective because it is a version of standard empiricism which, we have already seen, is untenable. It fails to identify the problematic aim of science properly, and thus fails to specify the need for science to improve its aims and methods as it proceeds.[10] Popper's notion of critical rationalism is defective in an analogous way. It does not make improving aims and methods, when aims are problematic, an essential aspect of rationality. It is a vital part of reason, but not the whole story.

If, however, we take the above aim-oriented empiricist conception of scientific method as our starting point, and generalize that, the outcome is different. It is not just in science that aims are problematic; this is the case in life too, either because different aims conflict, or because what we believe to be desirable and realizable lacks one or other of these features, or both. Above all, the aim of creating global civilization is inherently and profoundly problematic. Furthermore, it is not just science that suffers from rationalistic neurosis; many other institutional and traditional endeavours repress problematic aims and acknowledge ostensibly unproblematic, token, "neurotic" aims instead. Quite generally, then, and not just in science, whenever we pursue a problematic aim we need, first, to acknowledge the aim; then we need to represent it as a hierarchy of aims, from the specific and problematic at the bottom of the hierarchy, to the general and unproblematic at the top.

---

10. As I pointed out in note 6 of chapter one, early Popper (1959) defends "bare" standard empiricism (the doctrine that observation and experiment, plus the empirical content of theories, are the *only* considerations which determine choice of theory in science; later, Popper revised this, and required, in addition, that a new theory, in order to be acceptable, must "proceed from some *simple, new, and powerful unifying* idea", that is, it must satisfy a *"requirement of simplicity"* (Popper, 1963, p. 241), this being a version of "dressed" standard empiricism. What Popper cannot acknowledge is that persistently biasing choice of theory in the direction of simplicity and unity means that science makes a persistent, metaphysical assumption about the world. He cannot acknowledge this — the key step required to break with standard empiricism — because to do so would involve rejecting his principle of demarcation, which excludes metaphysical theses from scientific knowledge just because they are not empirically falsifiable.

In this way we provide ourselves with a framework within which we may improve more or less specific and problematic aims and methods as we proceed, learning from success and failure in practice what it is that is both of most value and realizable. Such an "aim-oriented" conception of rationality is the proper generalization of the aim-oriented, progress-achieving methods of science.

Any conception of rationality which systematically leads us astray must be defective. But any conception of rationality, such as Popper's critical rationalism, which does not include explicit instructions for the *improvement* of aims, must systematically lead us astray. It will do so whenever we fail to choose that aim that is in our best interests or, more seriously, whenever we misrepresent our aim — as we are likely to do whenever aims are problematic. In these circumstances, as we saw in chapter one, the more "rationally" we pursue the aim we acknowledge, the worse off we will be. Systematically, such conceptions of rationality, which do not include provisions for improving problematic aims, are a hindrance rather than a help; they are, in short, defective.

Aim-oriented empiricism and its generalization, aim-oriented rationality, incorporate all the good points of Popper's falsificationist conception of science and its generalization, critical rationalism, indicated above, but also improve on Popper's notions, in being designed to free science and other worthwhile endeavours of rationalistic neurosis. Popper's picture of science, despite all its good points, is still *neurotic* science, with the real, problematic aim repressed. And as a result, Popper's critical rationalism, despite its good points, is not designed to free endeavours with problematic aims of their neurosis; it is not designed to help improve aims and methods.[11]

So much for the second blunder, and how it is to be put right. We come now to the *third* blunder.

This is by far the most serious of the three blunders made by *The Traditional Enlightenment*. The basic Enlightenment idea, after all, is to learn from our solution to the first great problem of learning how to solve the second problem — to learn, that is, from scientific progress how to make social progress towards an enlightened world. Putting this idea into practice involves getting appropriately generalized progress-

---

11. For further details concerning aim-oriented rationality see (Maxwell, 1984a, especially chs. 4 and 5). For further details of the above critique of Popper see (Maxwell, 2002b, 2004a)

achieving methods of science *into social life itself!* It involves getting progress-achieving methods into our institutions and ways of life, into government, industry, agriculture, commerce, international relations, the media, the arts, education.   But in sharp contrast to all this, *The Traditional Enlightenment* has sought to apply generalized scientific method, not to social *life*, but merely to social *science!* Instead of helping humanity learn how to become more civilized by rational means, *The Traditional Enlightenment* has sought merely to help social scientists improve knowledge of social phenomena (this knowledge then being applied to help solve social problems).   The outcome is that today academic inquiry devotes itself to acquiring knowledge of natural and social phenomena, but does not attempt to help humanity learn how to become more civilized.   Instead of social inquiry having, as its basic task, to promote cooperatively rational tackling of problems of living in the social world, its primary task, rather, is to acquire knowledge of social phenomena.   Instead of being social methodology or social philosophy, social inquiry is pursued as social *science*.   This is the blunder that is at the root of our current failure to have solved the second great problem of learning.

As we shall see in a little more detail below, it is this third blunder that is primarily responsible for the severe rationalistic neurosis of social inquiry, and academic inquiry more generally.   For the outcome of the blunder is that social inquiry, and academic inquiry as a whole, devote themselves primarily to acquiring *knowledge*, rather than to helping humanity resolve its problems and conflicts of living in more cooperatively rational ways, thus acquiring *wisdom*.   The proper but profoundly problematic aim of promoting *wisdom* is repressed and replaced by the apparently much less problematic aim of acquiring *knowledge*. Wisdom is here to be understood as the capacity to realize what is of value in life for oneself and others, thus including knowledge, technological know-how and understanding, but much else besides.

Does Popper correct the third Enlightenment blunder? No, yes, and no. No, because this requires that the first and second blunders are corrected, and this Popper fails to do, as we have seen. Yes, because Popper applies critical rationalism (generalized from his falsificationist conception of scientific method) to life, to basic social and political problems: see above all his *Open Society and Its Enemies* (Popper,

1969). No, because when it comes to the nature of social inquiry, and academic inquiry more generally, Popper merely *perpetuates* the third blunder of *The Traditional Enlightenment*. For Popper, social inquiry is social *science*, with methods essentially the same as those of natural science.[12] And Popper never questions that the basic aim of academic inquiry is the pursuit of knowledge. In these respects, Popper is an 18th century *philosophe*.

In order to correct this third, monumental and disastrous blunder, we need, as a first step, to bring about a revolution in the nature of academic inquiry, beginning with social inquiry and the humanities. Social inquiry needs to be, not primarily social *science*, but rather social *methodology* or social *philosophy*, concerned to promote rational tackling of problems of living in the social world.[13] And the basic aim of academic inquiry needs to be, not just knowledge, but wisdom.

Let us now see, in a little more detail, what would result from correcting this third, monumental blunder of *The Traditional Enlightenment*. What we need to do is to see what results from applying the progress-achieving rules of reason (arrived at by generalizing the progress-achieving methods of science) to social *life* rather than to social *science*, to the task of making *social progress* towards a civilized world rather than to the task of making *intellectual progress* towards better knowledge of social phenomena. What we need to do, in other words, is see what would result from putting the following three steps of the Enlightenment programme, already indicated above, properly into practice:

(1) Specify correctly the progress-achieving methods of science.
(2) Generalize these methods so that they become fruitfully applicable to any worthwhile, problematic human endeavour, whatever the aims may be, and not just applicable to the scientific endeavour of

---

12. Popper decisively demolishes certain *conceptions* of social science, influential in the past, based on historicism (the doctrine that there are laws of historical development), but at the same time defends a highly traditional *neurotic* conception of social science, according to which natural and social sciences employ methods that are essentially the same (Popper, 1974, pp. 62-63 and 130-143): the point is discussed in (Maxwell, 1984a, pp. 195-198). Popper's critique of social science does not go nearly far enough.

13. Elsewhere I have spelled out in more detail than I am able to do here why this revolution is needed, and what it would amount to: see (Maxwell, 1984a). See also (Maxwell, 1976a; 1992a; and 2001c, ch. 9).

improving knowledge.

(3) Apply these generalized progress-achieving methods to the great human endeavour of trying to make social progress towards an enlightened, civilized world.

Step (1) of *The Enlightenment Programme* was discussed in chapters one and two; step (2) has been discussed above in the present chapter: our main concern, in what follows, is step (3).

I shall proceed in two stages. First, I will adopt a slightly improved version of Popper's conception of the rules of reason, which I shall call "problem-solving rationality"; second, I will adopt the further improved conception of aim-oriented rationality (arrived at by generalizing aim-oriented empiricism).

3.4 The New Enlightenment Employing Problem-Solving Rationality

Problem-solving rationality can be summed up in the following four rules of rational problem solving.

(1) Articulate, and try to improve the articulation of, the problem to be solved.
(2) Propose and critically assess possible solutions.
(3) When necessary, break up the basic problem to be solved into a number of preliminary, simpler, analogous, subordinate or specialized problems (to be tackled in accordance with rules 1 and 2) in an attempt to work gradually towards a solution to the basic problem to be solved.
(4) Interconnect attempts to solve basic and specialized problems, so that basic problem-solving may guide, and be guided by, specialized problem-solving.

Popper's critical rationalism consists of rules 1 and 2; problem-solving rationality improves on this by adding on rules 3 and 4, which become relevant when we are confronted by some especially recalcitrant problem — such as the problem of understanding the nature of the universe, or the problem of creating a civilized world — which can only be solved gradually and progressively, bit by bit, and not all at once. Popper was too hostile to specialization to emphasize the need for rule 3; he did not appreciate that the evils of specialization can be counteracted by implementing rule 4.[14]

It might seem that in moving from scientific method to critical and problem-solving rationality we lose the idea of learning from experience; but this is not so. Problem-solving rationality, as enshrined in the above four rules, is a method of learning from experience. Experience is what we acquire through trying out various possible solutions to the problem we wish to solve, and discovering that these possibilities more or less fail. Consider, for example, a problem of action, a technological or political problem, perhaps: in criticizing a proposed solution we may well appeal to the (adverse) outcome of attempting to put the solution into practice; that is, we appeal to experience. Experience, in this broad sense, is what we acquire through trying to do things, trying to solve problems: it is a generalization of the notion of experience as this arises in connection with science — observation and experimentation. Problem-solving rationality might also be called "problem-solving empiricism"; it is as much a generalization of scientific empiricism as it is of scientific rationality.

The above four rules, though by no means sufficient for rationality,[15]

---

14. See (Maxwell, 1980) for an account of problem-solving rationality, a critical discussion of mere specialization, and an account of Popper's views concerning specialization. I have presented problem-solving rationality as improving on Popper's critical rationalism in adding on rules (3) and (4). But it can also be seen as the outcome of generalizing an improved version of Popper's conception of scientific method, one which emphasizes the need for sustained attempts to articulate and solve fundamental metaphysical problems [rules (1) and (2)], in addition to the tackling of more specialized problems concerning empirically testable theories [rule (3)], there being sustained interplay between these two activities [in accordance with rule (4)]. This conception of scientific method is discussed in (Maxwell, 1980, pp. 42-49). It amounts to a version of aim-oriented empiricism, depicted in Fig. 1.2 of ch. 1.

15. These rules are not sufficient for rationality in part because of a lack of specific detail about how to improve aims and methods when aims are problematic, and in part because the list of rules is by no means complete. A very important additional rule is: 5. In seeking to solve a problem, P*, search for an analogous, already solved problem, P; if such a problem is found, modify the solution, S, appropriately, taking the similarities and differences between P and P* into account, so that S becomes S*, and consider this as a candidate solution to P*. For further details, see (Maxwell, 1984, chs. 4 and 5).

are certainly necessary for it. No mode of inquiry can hope to be rational which systematically violates any of these rules. In a moment we shall see that academic inquiry as it exists in the main at present, devoted to the pursuit of knowledge, systematically violates *three* of these four elementary, almost banal, entirely uncontroversial, rules of reason.

Two preliminary points now need to be made.

First, in order to create a more civilized, enlightened world, the problems that we need to solve are, fundamentally, problems of *living* rather than problems of *knowledge*. It is what we *do* (or refrain from doing) that matters, and not just what we *know*. Even where new knowledge or technology is needed, in connection with agriculture or medicine for example, it is always what this enables us to *do* that solves our problems of living.

Second, in order to make progress towards a sustainable, civilized world we need to learn how to resolve our conflicts in more cooperative ways than at present. A group acts cooperatively in so far as all members of the group share responsibility for what is done, and for deciding what is to be done, proposals for action, for resolution of problems and conflicts, being judged on their merits from the standpoint of the interests of the members of the group (or the group as a whole), there being no permanent leadership or delegation of power.[16] Competition is not opposed to cooperation if it proceeds within a framework of cooperation, as it does ideally within science. There are of course degrees of cooperativeness, from its absence, all out violence, at one extreme, through settling of conflicts by means of threat, manipulation, agreed procedures such as voting, via bargaining, to all out cooperativeness at the other extreme. If we are to develop a sustainable, civilized world we need to move progressively away from the violent end of this spectrum towards the cooperative end.[17]

---

16. As I am using the term, a conflict is only resolved "cooperatively" if it is resolved "justly".

17. All sorts of considerations render all-out cooperativeness in all aspects of life an ideal that is both undesirable and unrealizable. Aggression needs to be restrained with aggression; even limited cooperative action cannot exist without forceful restraint of those who would destroy cooperation; science would come to an end if everyone's views received equal attention; everyone benefits if those who are especially gifted, in the arts

Granted, then, that the task of academic inquiry is to put the four rules of problem-solving rationality into practice in such a way as to help humanity learn how to make progress towards a civilized, enlightened world, the primary intellectual tasks must be:

(1) To articulate, and try to improve the articulation of, those social problems of living we need to solve in order to make progress towards a better world.

(2) To propose and critically assess possible, and actual, increasingly cooperative social actions — these actions to be assessed for their capacity to resolve human problems and conflicts, thus enhancing the quality of human life.

These intellectually fundamental tasks are undertaken by social inquiry, at the heart of the academic enterprise. Social inquiry also has the task of promoting increasingly cooperatively rational tackling of problems of living in the social world — in such contexts as politics, commerce, international affairs, industry, agriculture, the media, the law, education.

Academic inquiry also needs, of course, to implement the third rule of rational problem-solving; that is, it needs:

(3) To break up the basic problems of living into preliminary, simpler, analogous, subordinate, specialized problems of knowledge and technology, in an attempt to work gradually towards solutions to the basic problems of living.

---

for example, have privileged opportunities to learn and practice their art, their gifts; cooperation without any delegation of responsibility to a few people will result, in all sorts of circumstances, from business to government, in hopeless inefficiency or outright chaos. The proper, problematic ideal is rational cooperation tempered by various other considerations of value and practicality. In the world as it is at present, riddled with armed conflict, actual and threatened, undemocratic governments, vast inequalities of wealth, violation of civil rights, gross persistent injustice, and the dominance of managerialism in the workplace, there is a great deal of room for an increase in cooperation of a desirable kind, before undesirable aspects of cooperation become apparent. It is vital to appreciate, however, that pursuing social goals in cooperatively rational ways needs to be *learned*: even where there is the best will in the world, still cooperative action requires skill and appropriate education if it is to meet with success.

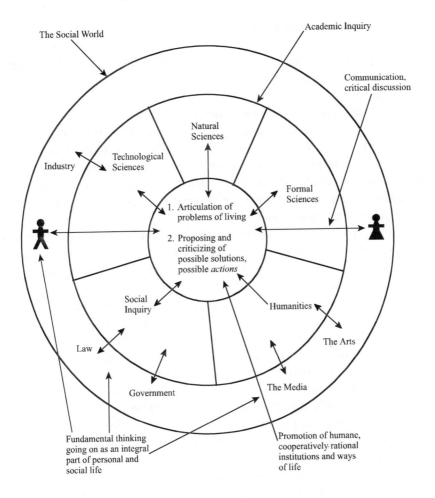

Figure 3.1: Wisdom-Inquiry Implementing Problem-Solving Rationality

But, in order to ensure that specialized and basic problem solving keep in contact with one another, the fourth rule of rational problem solving also needs to be implemented: that is, academic inquiry needs:

(4) To interconnect attempts to solve basic and specialized problems, so that basic problem-solving may guide, and be guided by,

specialized problem-solving.

In Fig. 3.1 I have tried to depict the kind of inquiry that would emerge as a result of putting the above four rules of rational problem-solving into academic practice, as just indicated.

I now spell out in a little more detail some of the basic features that academic inquiry would have were it to put this "problem-solving" conception of reason into practice. These features are direct consequences of taking seriously the basic idea that inquiry should use and be based on the above four rules of reason in seeking to help humanity learn how to solve its problems of living in increasingly cooperatively rational ways.

To begin with, social inquiry would not, primarily, be social science; it would have, rather, the intellectually basic task of engaging in, and promoting in the social world, increasingly cooperatively rational tackling of conflicts and problems of living. Social inquiry, so conceived, would be actually intellectually more fundamental than natural science (which would seek to solve subordinate problems of knowledge and understanding). Academic inquiry, in seeking to promote cooperatively rational problem solving in the social world, would engage in a two-way exchange of ideas, arguments, experiences and information with the social world. The thinking, the problem solving, that really matters, that is really fundamental, is the thinking that we engage in, individually, socially and institutionally, as we live; the whole of academic inquiry would be, in a sense, a specialized part of this, created in accordance with rule 3, but also being required to implement rule 4 (so that social and academic problem solving may influence each other). Academic inquiry, on this model, is a kind of peoples' civil service, doing openly for the public what actual civil services are supposed to do, in secret, for governments. Academic inquiry would need just sufficient power to retain its independence, to resist pressures from government, industry, the media, religious authorities, and public opinion, but no more. Academia would propose to, argue with, learn from, and would attempt to teach and criticize all sectors of the social world, but would not instruct or dictate. It would be an intellectual resource for the public, not an intellectual bully.

The basic intellectual aim of inquiry may be said to be, not knowledge, but wisdom — wisdom being understood to be the desire, the active endeavour and the capacity to realize[18] what is desirable and of value in life, for oneself and others. Wisdom includes knowledge, know-how and understanding but goes beyond them in also including the desire and active striving for what is of value, the ability to experience value, actually and potentially, in the circumstances of life, the capacity to help realize what is of value for oneself and others, the capacity to help solve those problems of living that need to be solved if what is of value is to be realized, the capacity to use and develop knowledge, technology and understanding as needed for the realization of value. Wisdom, like knowledge, can be conceived of not only in personal terms but also in institutional or social terms. Thus, the basic aim of academic inquiry, according to the view being indicated here, is to help us develop wiser ways of living, wiser institutions, customs and social relations, a wiser world.

So far, wisdom-inquiry (as it may be called) has been characterized as having the task of helping humanity learn how to tackle its problems of living more rationally; nothing has been said about learning from experience. But, as I indicated above, the four rules of reason that we are considering are also rules for learning from experience; this has a vital role to play in the conception of inquiry we are considering. What we learn as a result of attempting to put into practice some proposed solution to a problem of living is of course all important for learning how to build a better world. A vital task for academic inquiry (especially for history) is to monitor the successes and failures of our past attempts at solving problems of living. As far as possible we should try to ensure that our failed social experiments, our failed attempts at solving social problems, are performed only *in imagination*, and not *in practice* in the real world, so that we only suffer the consequences of failure in imagination, and not in reality. But however vivid, far-seeing and accurate our imagination may be, failure in practice will always happen, and we should seek to learn all we can from it for future actions. To this extent, the conception of inquiry we are considering can be regarded as a kind of empiricism. In two crucial respects, however, it differs from what is usually meant by empiricism. First, what is learned

---

18. "Realize" is intentionally ambiguous in that it here means *both* "to apprehend" and "to make real".

is how to do things, how to realize what is of value, how to live — that is, wisdom — and not, primarily, what we learn in the context of science: knowledge of fact. And secondly, as I have already remarked, "experience" means something like "what we acquire as a result of attempting to do things, attempting to realize what is of value", and not, primarily, what it means in the context of science: observation and experiment. (This latter meaning is a specialized version of the former meaning.)

It is important to appreciate that the conception of academic inquiry that we are considering is designed to help us to see, to know and to understand, for their own sake, just as much as it is designed to help us solve practical problems of living. It might seem that social inquiry, in articulating problems of living and proposing possible solutions, has only a severely practical purpose. But engaging in this intellectual activity of articulating personal and social problems of living is just what we need to do if we are to develop a good empathic or "personalistic" understanding of our fellow human beings (and of ourselves) — a kind of understanding that can do justice to our humanity, to what is of value, potentially and actually, in our lives. In order to understand another person *as a person* (as opposed to a biological or physical system) I need to be able, in imagination, to see, desire, fear, believe, experience and suffer what the other person sees, desires, etc. I need to be able, in imagination, to enter into the other person's world; that is, I need to be able to understand his problems of living as he understands them, and I need also, perhaps, to understand a more objective version of these problems. In giving intellectual priority to the tasks of articulating problems of living and exploring possible solutions, social inquiry thereby gives intellectual priority to the development of a kind of understanding that people can acquire of one another that is of great intrinsic value. In my view, indeed, personalistic understanding is essential to the development of our humanity, even to the development of consciousness. Our being able to understand each other in this way is also essential for cooperatively rational action.

And it is essential for science. It is only because scientists can enter imaginatively into each other's problems and research projects that objective scientific knowledge can develop. At least two rather different motives exist for trying to see the world as another sees it: one may seek to improve one's knowledge of the other person; or one may seek to improve one's knowledge of the world, it being possible that the other person has something to contribute to one's own knowledge. Scientific knowledge arises as a result of the latter use of personalistic understanding — scientific knowledge being, in part, the product of endless acts of personalistic understanding between scientists (with the personalistic element largely suppressed so that it becomes invisible). It is hardly too much to say that almost all that is of value in human life is based on personalistic understanding.[19]

The basic intellectual aim of the kind of inquiry we are considering is to devote reason to the discovery of what is of value in life. This immediately carries with it the consequence that the arts have a vital *rational* contribution to make to inquiry, as revelations of value, as imaginative explorations of possibilities, desirable or disastrous, or as vehicles for the criticism of fraudulent values through comedy, satire or tragedy. Literature and drama also have a rational role to play in enhancing our ability to understand others personalistically, as a result of identifying imaginatively with fictional characters — literature in this respect merging into biography, documentary and history. Literary criticism bridges the gap between literature and social inquiry, and is more concerned with the *content* of literature than the means by which it achieves its effects.

Another important consequence flows from the point that the basic aim of inquiry is to help us discover what is of value, namely that our feelings and desires have a vital rational role to play within the intellectual domain of inquiry. If we are to discover for ourselves what is of value, then we must attend to our feelings and desires. But not everything that feels good is good, and not everything that we desire is desirable. Rationality requires that feelings and desires take fact, knowledge and logic into account, just as it requires that priorities for scientific research take feelings and desires into account. In insisting on this kind of interplay between feelings and desires on the one hand, knowledge and understanding on the other, the conception of inquiry that we are considering resolves the conflict between rationalism and

---

19. For a more detailed discussion of the nature, significance, and intellectually fundamental character of "personalistic" understanding, and its role in "wisdom-inquiry", see (Maxwell, 1984a, pp 172-89 and 264-75; 2001c, chs. 5-7).

romanticism, and helps us to acquire what we need if we are to contribute to building civilization: mindful hearts and heartfelt minds.

This, then, in bare outline, is the kind of academic inquiry that would have emerged from the 18th century Enlightenment if the third great blunder of the Enlightenment had not been made.

But the blunder was made. Instead of the progress-achieving methods of science (generalized to become problem-solving rationality) being applied to *social life*, scientific method was applied to the task of developing *social science* alongside natural science. The outcome is what we have (by and large) today: a kind of inquiry — which we may call knowledge-inquiry to distinguish it from wisdom-inquiry — that gives intellectual priority to the task of acquiring *knowledge*, this knowledge, once acquired, being subsequently and secondarily applied to help solve social problems. Rule 3 of problem-solving rationality is put into practice to splendid effect: the outcome is the maze of specialized disciplines of the formal, natural, social and technological sciences that go to make up much of academic inquiry today. But rules 1, 2 and 4 are violated. Academic inquiry today, restricted primarily to solving problems of knowledge, is so irrational, in a wholesale and structural way, that *three* of the four most elementary rules of reason conceivable are violated. Rule 1 is violated because academia can articulate problems of *knowledge* but cannot, at a fundamental level, articulate problems of *living*. Rule 2 is violated because academia can propose and critically assess possible solutions to problems of *knowledge* — theories, observational and experimental results, factual claims of all kinds — but cannot propose and critically assess possible solutions to problems of living — proposals for action, policies, political programmes, political philosophies, philosophies of life. All these latter do not state matters of fact; they embody proposals as to what we should do, how we should live, what we should seek to change and create; they incorporate such things as values, human hopes and fears, policies, strategies for living: they do not constitute potential contributions to knowledge, and are thus excluded from a kind of inquiry devoted to the pursuit of knowledge. Once rules 1 and 2 are violated, rule 4 is necessarily violated as well.[20]

---

20. Discussion of problems of living and how to solve them goes on at present at the fringes of academic inquiry, within such disciplines as peace studies, development studies, social policy studies, medicine, agricultural science and other applied sciences.

This wholesale, structural irrationality of academic inquiry as it mostly exists today is no mere formal matter. It has far-flung, long-term damaging consequences.

It means we lack what at present we most need: sustained, intelligent, imaginative, unconstrained exploration of our local and global problems of living and what we might do to help solve them, carried on in a public, influential manner. Problems of third world poverty, vast inequalities of wealth across the globe, environmental problems, problems of finite resources such as oil, water, land, minerals, problems of population growth, problems of war, persistent violence, terrorism, dictatorships, and extreme violation of civil rights, problems and threats posed by the arms trade and the spread of modern armaments, whether conventional, chemical, biological or nuclear: it is not just that wealthy, democratic nations fail to act appropriately in response to these grim, endemic problems; there is not even good, sustained, public discussion about what should be done in response to them. Academia cannot supply this, being restricted (primarily) to the pursuit of knowledge; the public pronouncements of politicians, journalists, religious leaders and others are no substitute. Those who do have something valuable to say find themselves powerless to get their message across.

This absence of sustained exploration of our problems of living means, furthermore, that knowledge and technological know-how are pursued in a way that is dissociated intellectually from a more fundamental concern to promote increasingly cooperatively rational tackling of conflicts and problems of living. As I have pointed out above, it is this that is at the root of most of our current global problems.

Knowledge-inquiry, a kind of inquiry that pursues knowledge and technological know-how, and fails to give intellectual priority to the tasks of articulating our problems of living, and proposing and criticizing possible solutions (thus violating three of the four most elementary rules of reason conceivable), must inevitably tend to *create* the kind of global problems we face today, the outcome of possessing much recently acquired power to act without the power to act wisely. And the more successful such knowledge-inquiry is, so the greater the human suffering

---

The crucial point is that such discussion is at the periphery; it is not intellectually central and fundamental.

it is likely to lead to. Reason is far too important "for everything we cherish" for it to be tolerable that it should be systematically violated in this way.

3.5 The New Enlightenment Employing Aim-Oriented Rationality

"Problem-solving rationality", though an essential part of reason, is nevertheless defective. As we have seen, it does not explicitly help improve aims when these are problematic, and is not designed to cure rationalistic neurosis.[21] Let us now see what kind of social inquiry, and academic inquiry more generally, would emerge if aim-oriented rationality is appealed to, a conception of reason that is designed quite specifically to improve problematic aims, and cure rationalistic neurosis. This version of *The New Enlightenment* incorporates everything of value in the kind of inquiry just depicted, and develops it further.

We are to assume, then, that a basic task of academic inquiry is to help humanity gradually get more aim-oriented rationality into diverse aspects of social and institutional life — personal, political, economic, industrial, agricultural, educational, international — so that humanity may gradually learn how to make progress towards an enlightened world. Social inquiry, in taking up this task, needs to be pursued as social *methodology* or social *philosophy*. What the philosophy of science is to science, as conceived by aim-oriented empiricism, so sociology is to the social world: it has the task of helping diverse valuable human endeavours and institutions gradually improve aims and methods so that the world may make social progress towards global enlightenment. (The sociology of science, as a special case, is one and

---

21. Problem-solving rationality and aim-oriented rationality are nevertheless intimately connected. Any problem can be construed to consist of an aim, some kind of tentative route to the realization of the aim, and a barrier which prevents the route from being successfully pursued. On the other hand, all aim-pursuing consists of problem-solving. When we pursue aims or perform actions instinctively, without difficulty, we may not even notice the incredibly sophisticated problem-solving that our brains perform in guiding our actions, but these would be very apparent to artificial intelligence experts if they attempted to create a robot capable of mimicking our actions.

The nearest that problem-solving rationality comes to stipulating that one should try to improve problematic aims comes in rule 1 "Articulate, *and try to improve the articulation of*, the problem to be solved". This might include improving the aim implicit in the problem to be solved.

the same thing as the philosophy of science, as we shall see below.) And a basic task of academic inquiry, more generally, becomes to help humanity solve its problems of living in increasingly rational, cooperative, enlightened ways, thus helping humanity become more civilized. The basic aim of academic inquiry becomes, I have already said, to promote the growth of *wisdom*. Those parts of academic inquiry devoted to improving knowledge, understanding and technological know-how are pursued in such a way as to contribute to the growth of wisdom.

As I have already remarked, the aim of achieving global civilization is inherently problematic.[22] This means, according to aim-oriented rationality, that we need to represent the aim at a number of levels, from the specific and highly problematic to the unspecific and unproblematic.

Thus, at a fairly specific level, we might, for example, specify civilization to be a state of affairs in which there is an end to war, dictatorships, population growth, extreme inequalities of wealth, and the

---

22. Fundamentally, this is due to the profound difficulty of discovering what is *achievable* (by increasingly civilized means), and *of value*. People hold conflicting views about what is achievable and of value, all too often in a highly dogmatic way, ignoring the profoundly problematic character of the whole idea of civilization. Many well-known views that have been proposed as to what constitutes Utopia, an ideally civilized society, have been unrealizable, horrifically undesirable, or both, attempts to realize such ideals, when taken up in practice, leading to various kinds of hell on earth (as in Hitler's Germany, Stalin's Russia, or Mao's China). Furthermore, it is not just that people have conflicting interests, values and ideals; even our very best ideas as to what constitutes civilization embody (and need to embody) conflicting ideals. Thus freedom and equality, even though inter-related, may nevertheless clash. It would be an odd notion of individual freedom which held that freedom was for some, and not for others; and yet if equality is pursued too single-mindedly this will undermine individual freedom, and will even undermine equality, in that a privileged class will be required to enforce equality on the rest, as in the Soviet Union before its collapse. A basic aim of legislation for civilization, we may well hold, ought to be to increase freedom by restricting it: this brings out the inherently problematic character of the aim of achieving civilization. One thinker who has stressed the inherently contradictory character of the idea of civilization is Isaiah Berlin: see, for example, (Berlin, 1980, pp. 74-79). Berlin thought the problem could not be solved, but this was because he was ignorant of aim-oriented rationality. In depicting ideals of civilization at a hierarchy of levels, aim-oriented rationality provides the means for progressively improving resolutions to inherently conflicting ideals, such as freedom and equality.

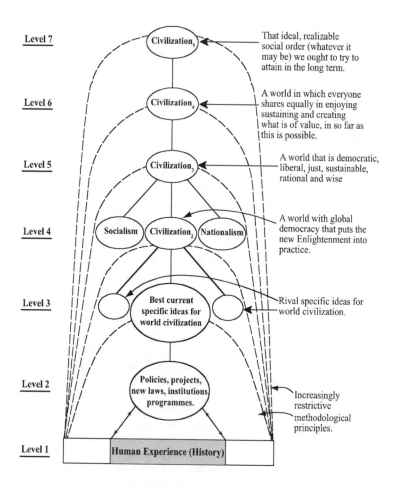

Figure 3.2: Implementing Aim-Oriented Rationality in Pursuit of Civilization

establishment of democratic, liberal world government and a sustainable world industry and agriculture. At a rather more general level we might specify civilization to be a state of affairs in which everyone shares equally in enjoying, sustaining and creating what is of value in life *in so far as this is possible.* Figure 3.2 depicts a cartoon version of what is required, arrived at by generalizing and then reinterpreting Fig. 1.2.

In addition, specific institutions, such as government, industry, agriculture, commerce, the law, the media and education, need to acquire the hierarchical structure of aims and methods of aim-oriented rationality, analogous to the aim-oriented empiricist structure of science: see Fig. 3.3. The *philosophy* of government, law, education or whatever it may be — that is the imaginative and critical exploration of actual and possible aims and methods — needs to be carried on in such a way that it is capable of influencing, and being influenced by, the institution, the human endeavour, it is the philosophy *of*. Philosophy, in this sense, serves a severely practical purpose. A basic and highly problematic requirement of this kind of practical, empirical philosophy is that it is pursued in such a way that (a) it is open to everyone, especially those outside the institution in question, so that the self-serving bias institutions tend to acquire can be corrected, and (b) only the best ideas come to the fore and become influential over what goes on in practice.

As a result of building into our institutions and social life such a hierarchical structure of aims and associated methods, we create a framework within which it becomes possible for us progressively to improve our real-life aims and methods in increasingly cooperative ways as we live. Diverse philosophies of life — diverse religious, political, economic and moral views — may be cooperatively developed, assessed and tested against the experience of personal and social life. It becomes possible progressively to improve diverse *philosophies of life* (diverse views about what is of value in life and how it is to be realized) much as *theories* are progressively and cooperatively improved in science. In doing this, humanity would at last have learned from the solution to the first great problem of learning how to go about solving the second problem.

Aim-oriented rationality is especially relevant when it comes to resolving conflicts cooperatively. If two groups have partly conflicting aims but wish to discover the best resolution of the conflict, aim-oriented rationality helps in requiring of those involved that they represent aims at a level of sufficient imprecision for agreement to be possible, thus creating an agreed framework within which disagreements may be explored and resolved. Aim-oriented rationality cannot, of itself, combat non-cooperativeness, or induce a desire for cooperativeness; it can however facilitate the cooperative resolution of conflicts if the desire for this exists. In facilitating the cooperative resolution of conflicts in this way, aim-oriented rationality can, in the long term, encourage the desire

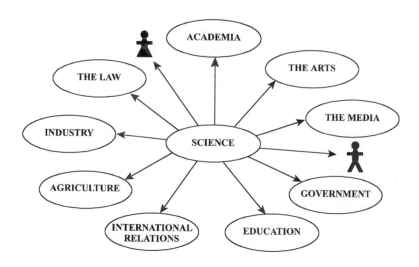

Figure 3.3: The Enlightenment Programme: Extracting Progress-Achieving Methods from Science, Generalizing Them, and Applying Them to Other Institutions and Aspects of Social Life

for cooperation to grow (if only because it encourages belief in the possibility of cooperation).

Aim-oriented rationality synthesizes traditional Rationalist and Romantic ideals of intellectual integrity in that it requires *both* attention to objective fact, logic, science, *and* emotional and motivational honesty, honesty about aims.

Wisdom-inquiry can be seen as a much more natural development of animal learning than can knowledge-inquiry. For animal learning (like wisdom-inquiry) is concerned primarily with learning how to solve problems of *living*, problems of *action*. Animals acquire knowledge about their environment in a way that is subordinated to the primary task of discovering how to solve problems of living — to find something to eat, to avoid a predator, to find a mate, to care for offspring. (Some animals do, however, exhibit curiosity about novelties in their environment.) The big differences between animal learning and wisdom-inquiry are the far more pronounced social and cultural aspects of the latter, and the fact that, whereas the former has, as its basic aim, to promote survival and reproductive success, the latter has, as its basic *problematic* aim, to promote life *of value*. Evolution equips our brains to work out how to attain the predetermined goals of survival and reproductive success, but does not equip our brains to discover how to *improve* our basic aims in life as we live, so that we come to realize what is of most value, potentially, in the circumstances in which we find ourselves (especially as the world we now live in is so different from the world in which humanity first evolved). This may explain in part why it is so difficult for us to put aim-oriented rationality into practice in life, and also may serve to emphasize how important it is for us to try to do it.[23]

3.6 The Damaging Neurosis of Social Science

The rationalistic neurosis of natural science has not inhibited scientific progress too much, not, at least, if the aim of science is taken to be to improve expert knowledge.[24] This is because dressed standard empiricism, put into scientific practice in a sufficiently unrigorous way, does not differ too much from aim-oriented empiricism.[25] It is the philosophy of science, the neurotic face of science, engaged in the fruitless task of rationalization, of trying to justify the unjustifiable neurotic view of standard empiricism, that really suffers. Fortunately, scientists have not taken the philosophy of science too seriously.

---

23. For a more detailed discussion of the evolutionary roots of human inquiry, and the importance of setting human inquiry into an evolutionary context, see (Maxwell, 1984a, p. 116, pp. 174-181, 193-195 and 267-275: 2001c, chs. 7 and 9).

24. The neurosis of natural science is rather more serious, however, if one takes the aim of science to be to contribute to the enrichment of human life, by both cultural and technological means. Science has been extraordinarily unsuccessful in communicating to the public at large the overall view of the universe that emerges from scientific inquiry: and, as we have seen, the technological blessings of science are mixed. These failings of science to help enrich human life are in part due to the neuroses of science, already discussed.

25. If standard empiricism were to be implemented rigorously, in such a way that science makes no persistent (possibly implicit) metaphysical assumption about the universe, then science would come to a standstill. Science would be overwhelmed by endlessly many empirically successful disunified theories. If implemented unrigorously, as at present, so that preference is given to unifying theories even against the evidence (and a persistent, implicit metaphysical assumption is made), science is still adversely affected, to a degree, as we saw in chapter two.

When we come to social inquiry, however, all this changes dramatically. Here, rationalistic neurosis really does matter, and has far-reaching, long-term damaging consequences, both intellectual and humanitarian.

The rationalistic neurosis of social inquiry amounts to this. The social sciences arose and were developed in response to the Enlightenment idea of learning from scientific progress how to achieve social progress towards an enlightened world. The proper, fundamental aim of social inquiry is to help humanity learn how to become enlightened by cooperatively rational means. The *philosophes* of the Enlightenment, and especially those who came after them, thought that this meant developing social inquiry as social *science*: first, knowledge of society is to be developed; then, once acquired, it can be applied to help solve social problems. But this, as we have seen, is to commit a disastrous blunder: in order to implement the basic Enlightenment idea properly, social inquiry needs to be social *methodology*, or social *philosophy*, not primarily social *science*. The proper, basic task of social inquiry is to get into our diverse institutions, traditions and ways of life, into the fabric of society, general progress-achieving methods arrived at by generalizing the progress-achieving methods of science. The *neurotic* aim is to restrict the task of social inquiry to acquiring knowledge of social phenomena; the real, un-neurotic aim is to help humanity tackle its problems of living by increasingly cooperatively rational means. The difference between the current neurotic character of social inquiry as social *science*, and the proper, un-neurotic character of social inquiry as the promotion of cooperatively rational tackling of problems and conflicts and living, and as social *methodology*, is dramatic and profound.

Furthermore, the rationalistic neurosis of social inquiry has far-reaching damaging consequences. It is this which ensures that humanity has so far *failed* to learn from its solution to the first great problem of learning how to solve the second great problem of learning. It is this which has prevented us from developing a kind of inquiry that is well-designed from the standpoint of helping us gradually learn how to become civilized.

The neurosis of natural science may not matter too much, for natural science itself, at least. But the (associated) neurosis of social inquiry is a disaster, both for social inquiry itself, and for humanity.

We need, then, a revolution throughout the diverse branches of social inquiry — economics, sociology, anthropology, psychology, history, political science. These disciplines are not primarily *sciences*; they are not, primarily, concerned to improve knowledge of social phenomena: their primary task is to help us tackle our problems of living by increasingly cooperatively rational means so that we may gradually make progress towards a civilized world. Each discipline, of course, seeks to acquire knowledge about people, cultures, institutions and social structures where this helps us discover what our problems of living are, and what we might be able to do about them. The primary intellectual task, however, is to promote increasingly cooperatively rational tackling of problems of living; it is not to acquire *knowledge* of social phenomena. Economics is concerned with the economic aspect of our problems, psychology with the personal, the psychological aspect; history and anthropology seek to inform us about our past successes and failures in tackling problems of living — a record of past successes and failures being essential to rationality. Sociology tries, amongst other things, to help us build aim-oriented rationality into our various institutions and social endeavours: politics, industry, agriculture, the law, the media, international relations, education, the arts. The *sociology* of theatre, for example, is the *philosophy* of theatre, the exploration and critical assessment of rival views as to what the aims and methods of theatre ought to be.[26]

3.7 Philosophy and Sociology of Science

This revolution in social inquiry, necessary if social inquiry is to free itself of its neurosis and become fully beneficial to humanity, has one amusing consequence. It means that the *philosophy of science* becomes one and the same thing as the *sociology of science*.

At present, these two disciplines are so different, so much at odds with each other, that they scarcely speak to each other. Each is obliged to exist within the current framework, the current "paradigm" according to which the overall intellectual aim of academic inquiry is to acquire *knowledge*. The philosophy of science struggles to solve the misconceived, neurotic problems thrown up by standard empiricism. It

---

26. For further details see (Maxwell, 1984a).

is normative in character; it seeks to formulate the rules, the methods, that science *ought* to employ in order to meet with success: but unfortunately, nothing that it comes up with seems to be of much use to science itself (as is, of course, to be expected, given the neurotic character of the discipline). The sociology of science, on the other hand, seeks to acquire *sociological* knowledge about science: it sees itself as a part of sociology, in turn a part of social science. The sociology of science is thus factual, not normative: it scorns the normative task of explicating methods that science *ought* to implement. Each discipline is more or less contemptuous of the work of the other.[27]

Free the natural and social sciences of their neuroses, however, in the ways already indicated, and it is at once clear that the philosophy and sociology of science are one and the same discipline. The philosophy of science, in freeing itself from the hopeless, neurotic task of trying to prop up standard empiricism, needs to consider aims of science that are not just narrowly conceived intellectual aims; it needs to consider broader, and more fundamental human, humanitarian or social aims. In doing so, philosophy of science needs to come to grips with the institutional and social structure and character of science. The sociology of science on the other hand, in freeing itself of the general neurosis of social science, becomes social methodology, or social philosophy, concerned to explore and critically assess possible and actual aims and methods for science: this is, of course, the philosophy of science. The new synthesized discipline, philosophy-sociology of science, needs to be pursued as an integral part of science itself, influencing and being influenced by scientific practice, as required by aim-oriented empiricism and rationalism. At the same time non-scientists, members of the public, need to be able to contribute to the discipline.

In general, we may say that what the philosophy/sociology of science is to science, so the sociology of politics (let us say) is to politics itself. Social inquiry, quite generally, is to society what un-neurotic philosophy/sociology of science is to science.

---

27. For sociologists' perspectives see (Bloor, 1991; Barnes, 1974; Barnes, Bloor and Henry, 1996). For philosophers' perspectives see (Laudan, 1977, ch. 7; Newton-Smith, 1981, ch. 10; Brown, 2001). See also (Fuller, 1993; Segerstrale, 2000).

3.8 Academic Neurosis

Not only are natural science and social inquiry neurotic; academic inquiry taken as a whole suffers from rationalistic neurosis. As I have argued above, the fundamental intellectual and social aim of academic inquiry (science, social inquiry, the humanities, technological research, the formal sciences, education) ought to be to promote *wisdom*, where wisdom is defined, as I have already indicated, as the capacity to realize what is of value in life, for oneself and others.[28] Wisdom, so defined, includes knowledge, understanding and technological know-how, but much else besides. Like knowledge, wisdom can be conceived of as something that individual persons possess, and can also be conceived of in more impersonal terms as something possessed by institutions.

Inquiry rationally devoted to promoting wisdom — to helping humanity learn how to make progress towards a wise world — is what would have emerged if the 18th century Enlightenment programme had been developed and implemented free of blunders, if, in particular, social inquiry had been developed as social *methodology* rather than social *science*. If this had happened, we would have solved the second great problem of learning — the problem of learning how to become civilized or wise. But this did not happen. Instead, as I have already emphasized, social inquiry was developed as social science, and as a result what we have at present is a kind of academic inquiry which, though ultimately devoted to promoting human welfare (at least in principle) takes as its primary intellectual goal the pursuit of knowledge and technological know-how. This is the neurotic aim of academic inquiry. And this neurosis is, again, profoundly damaging in that it is this which has sabotaged the efforts of humanity to solve the second great problem of learning, to such an extent that most people today probably believe that the problem is permanently insoluble.

3.9 Philosophical Neurosis

It may seem almost beyond belief that intellectual inquiry is riddled with rationalistic neurosis in the way that I have indicated. One can

---

28. See also (Maxwell, 1984a, 1992a, 2001c, ch. 9).

perhaps understand the pressures which led to such a state of affairs arising in the first place, centuries ago. In Europe in the 15th and 16th centuries, speculating about the nature of the universe could result in one being burnt at the stake. In the 18th century one risked imprisonment and worse if one speculated about such things as freedom, justice and democracy, and criticized existing religious and secular authorities. This continues to be the case in many parts of the world up to the present (2004). It is entirely understandable that there are immense pressures on academics to refrain from tackling awkward problems experienced by people in their lives, having to do with such things as poverty, injustice, tyranny, bigotry, enslavement. But what is astonishing is that a certain wholesale avoidance of tackling such difficult problems of living in favour of tackling much less explosive problems of knowledge should persist even though this flies in the face of reason in the ways indicated. The neurotic blunders of natural science, social inquiry, and academic inquiry as a whole are *philosophical* blunders, in that they are blunders about fundamental *aims and methods*. Where have the philosophers been all this time? Why have they failed to point these blunders out?

The answer is that philosophy is, perhaps, the most neurotic discipline of all. Far from struggling to help free other disciplines from their neurotic straightjackets, philosophy has been most tightly bound in its own neurotic constraints.

The proper, un-neurotic aim of philosophy is to help solve the most general, the most fundamental problems that there are. One absolutely fundamental, general problem is what may be termed "the human world/physical universe problem". This is the problem of understanding how the world as we experience it, imbued with sensory qualities, consciousness, free will, meaning and value, can be imbedded in the physical universe as conceived of by modern physical science.[29] (We

---

29. Cartesian dualism is, of course, an early and immensely influential attempt at a solution to this problem. Much subsequent philosophy has struggled with problems engendered by Cartesian dualism: the mind-body problem, the problem of our knowledge of the external world, the problem of free will. But ironically, even though Cartesian dualism is nowadays rejected by most philosophers, this has not resulted in philosophy recognizing as fundamental to the discipline the problem which Cartesian dualism attempts to solve (or should be interpreted as attempting to solve). It is rare to find a philosopher asserting that the human world/physical universe problem is the fundamental problem of the discipline. Introductory courses in philosophy do not take this to be the basic problem of philosophy. In so far as such a tendency does exist, it began, perhaps,

encountered this problem in chapter one, as arising in a particularly acute form once the un-neurotic aim of physics of improving knowledge and understanding of the universe *presupposed to be physically comprehensible* is acknowledged.) The first serious task for philosophy is to keep alive an awareness of this problem, and keep alive imaginative and critical attempts to solve the problem. Philosophy (until recently, perhaps) has singularly failed to do this.

A second, related task for philosophy is to help humanity improve the aims and methods of various worthwhile endeavours in the light of problems (thus putting aim-oriented rationality into practice at a fundamental level). This includes the task of freeing inquiry from its neuroses. Again, philosophy has singularly failed to engage in this vital, fundamental task.

Instead, noble exceptions aside, philosophy in the 20th century has been split between arrogant, unintelligible bombast on the one hand, and sterile conceptual analysis on the other. On the one hand, philosophers like Bradley, McTaggart and Heidegger have over-reached themselves, and have claimed to be able to arrive at secure knowledge about ultimate realities independently of science. On the other hand, other philosophers, most notably G.E. Moore, Bertrand Russell (in some of his phases) and the logical positivists, reacting against such bombast, have insisted on only an extremely modest role for philosophy, to the point almost of intellectual self-annihilation.

Thus, the logical positivists divided up all meaningful propositions into two categories, the empirical and the analytic. The empirical are verified by an appeal to experience, while the analytic are verified by an appeal to the meaning of constituent terms. "The earth has at least one moon" is empirical because it can be established to be true by observation, while "All bachelors are unmarried" is established to be true by an appeal to the meaning of "unmarried" and "bachelor". Science deals with the empirical; philosophy, not being an empirical science, must be devoted to establishing analytic propositions by means of analysis of meaning.

This positivist view was long ago rejected by almost everyone for all

---

with (Smart, 1963). The thesis that the human world/physical universe problem is fundamental to philosophy is defended in some detail in (Maxwell, 2001c).

sorts of reasons, some good, many bad.[30] Mysteriously, however, the impoverishing implications for philosophy were not. For decades after logical positivism was first put forward in the 1930s, philosophy "in the analytic tradition" restricted itself to the analysis of concepts: mind, matter, truth, knowledge, justice, reality, and so on.[31] But philosophy conducted in this way, as analysis of concepts is, in the first place, intellectually dishonest. What philosophical concepts mean usually depends on what factual or evaluative ideas — what philosophical doctrines — are lurking behind the concepts: under the guise of mere analysis, philosophers surreptitiously insinuate unacknowledged, substantial philosophical doctrines, as when Ryle insinuates behaviourism via his "analysis" of mental concepts (Ryle, 1949), and at the same time denies he is doing any such thing. This way of doing philosophy is also a recipe for intellectual sterility. Developing new ideas (in an attempt to solve substantial problems) often involves disrupting established, philosophically "analysed" concepts: stick with such concepts, and you make it all but impossible to develop new ideas, think new thoughts. One might think that conceptual analysis is, these days, a thing of the past; but no, the "analytic" way of doing philosophy still lingers on. Philosophy "in the analytic tradition" has only very slowly and partially recovered from this travesty of what philosophy

---

30. The chief mistake in the positivists' thesis is the claim that all factual assertions are *empirical*, that is, observationally verifiable or falsifiable. This implies, as the positivists intended, that there can be no such thing as meaningful, factual *metaphysical* assertions — assertions that are neither verifiable nor falsifiable. But, as we have already seen, not only are such assertions meaningful; not only can they make substantial factual claims about the world: in addition, they can play an important role in science. An example is the (level 4) thesis that the universe has a unified dynamic structure (i.e. is physically comprehensible). This thesis is actually an important component of scientific knowledge.

There is, however, a correct and important point lurking within the incorrect thesis of the positivists. It is that no factual assertion about the world is necessary; all necessary assertions are devoid of factual content (in that they are assertions like "all bachelors are unmarried" or "2+2=4", true by definition). This, crucially, denies that there can be assertions that are both metaphysical and necessary — assertions that make substantial assertions about the world which, at the same time, can be proven like the theorems of pure mathematics. Not only did the logical positivists fail to articulate this point correctly; they were not even the first to make the point. It was made two centuries before the positivists by David Hume.

31. Three influential "classics" of neurotic conceptual analysis are: (Wittgenstein, 1953; Ryle, 1949, and Austin, 1962).

ought to be — this extreme philosophical neurosis.

Suffering from this self-imposed neurosis, analytic academic philosophy has been quite unable to perform its proper, serious, non-neurotic task: to tackle rationally (i.e. imaginatively and critically) our most general and fundamental problems, including problems concerning fundamental aims and methods of science, and of academic inquiry more generally. This is, of course, the worst failing of analytic philosophy.

A part of the problem is that, ever since logical positivism, analytic academic philosophers have been anxious about where and how philosophy fits in, as a respectable academic speciality alongside other specialities. (It was one of Wittgenstein's aims to turn philosophy into a respectable speciality, with an established method for analyzing meanings.) In the old days, philosophy, next to theology, was the supreme discipline. Then branches of philosophy split off and became established as independent sciences: first, natural philosophy became science; then disciplines such as economics and psychology split off from philosophy and became social sciences; and more recently, cosmology, logic and linguistics ceased to be a part of philosophy, and became independently established sciences. It began to seem that nothing could be left after this process of attrition. Philosophy had become vacuous.

But this traditional view of philosophy spawning scientific disciplines until nothing remains of itself rests on an untenable conception of science, and its relationship with philosophy, as we have seen. Granted standard empiricism, philosophy has no role within science. But granted aim-oriented empiricism, philosophical, or metaphysical assumptions constitute a vital, central part of scientific knowledge. Furthermore, the philosophical task of developing and critically assessing the aims and methods of science, and of inquiry more generally, is vital for science, and for inquiry as a whole. Failure to engage in these tasks, due to neurotic obsession with "analysis", has made it possible for the neurosis of science, and of academic inquiry, to perpetuate themselves unnoticed.

It is vital to appreciate, more generally, that not all academic specialities fit together in the same way. Philosophy is not just a speciality alongside other specialities. It is concerned with our most fundamental, general problems; the chief task of professional philosophers is to provoke others into thinking seriously about fundamental problems. It is not to do philosophy for others but, on the contrary, to keep alive a general discussion of philosophical problems above all amongst non-philosophers. A philosopher must be a kind of professional intellectual dilettante, interested in everything *except* academic philosophy, as it exists at present.[32]

In qualification of the above, I must add that in the last decade or so, in my view, there has been an upsurge in the intellectual vitality and seriousness of academic philosophy that is nominally "in the analytic tradition", due precisely to the repudiation of the idea that philosophy should restrict itself to analysis of concepts, and to a return to the idea that philosophy should tackle real, fundamental problems about the real world. This may be due, in part, to the influence of Karl Popper. Not only has Popper vehemently opposed the view that philosophy should devote itself to conceptual analysis; not only has he passionately affirmed the idea that philosophy should be about real problems with their roots in science, politics, the arts, life (see especially Popper, 1963, chs. 2 and 5; and Popper, 1976). In addition, he has practised what he has preached, as I tried to indicate above, in section 2, in discussing Popper's contributions to the enlightenment programme of learning from scientific progress how to achieve social progress.[33] Popper's *The Logic of Scientific Discovery* (first published in German in 1934) has had a profound impact, not just on the philosophy of science, but on science itself, on how scientists see the scientific enterprise. And Popper's *The Open Society and Its Enemies* has been immensely influential in such diverse fields as politics, economics, studies of Plato, ancient Greek philosophy and mathematics, Marxist thought, history of art, social and political philosophy. The book contains the classic refutation of Marxism and, together with George Orwell's *Animal Farm*, had an important role in discrediting Marxist thought when circulated in samizdat form in the Soviet bloc during the Cold War.[34]

---

32. For a development of this theme see (Maxwell, 1980).

33. For a recent critical assessment of Popper's work, see (Maxwell, 2002b).

34. One can indicate how Popper's contribution has helped to liberate philosophy from the stultifying influence of logical positivism, and the resulting neurotic activity of "conceptual analysis" as follows. As long as "knowledge" means "verified knowledge", then philosophy, as a body of untestable speculations about how to solve fundamental problems, has no role to play in contributing to knowledge. Furthermore, the impossibility of arriving at secure knowledge about the world by means of pure thought alone (there being no necessary metaphysical theses, as Hume understood, and as the

Another philosopher who has, in my view, helped to revitalize philosophy is J.J.C. Smart. His *Philosophy and Scientific Realism* (Smart, 1963), tackled in a fresh and lively way a fundamental problem of philosophy: how can we understand the world as we experience it given what modern science seems to be telling us about the world and ourselves? The brain process theory of consciousness, developed by Place (1956), Smart (1963), Armstrong (1968) and others, has had an immensely fruitful impact on subsequent philosophical work on such problematic issues as consciousness and free will. Philosophers working on these problems have been forced to take note of developments in neighbouring fields of study, in psychology, artificial intelligence, neuroscience, evolutionary theory, biology, and even physics. Three excellent works of philosophy, published since 1995, that illustrate this fruitful development, are: Daniel Dennett's *Darwin's Dangerous Idea* (1996), David Chalmers' *The Conscious Mind* (1996), and Robert Kane's *The Significance of Free Will* (1996).

Furthermore, many philosophers these days tackle such real-life issues as animal rights, feminism, racism, globalization, environmental degradation, problems of war, terrorism, development, tyranny: all this may be held to constitute the first steps in putting wisdom-inquiry into practice: see for example (Nagel, 1979; Singer, 1995; 2002; Regan, 2004).

Despite these encouraging developments, philosophy does not yet have the confidence to explore and critically assess actual and possible

---

logical positivists tried, but failed, to say), means that philosophy cannot deliver such knowledge, by thought alone. But accept Popper's theses that (1) theoretical scientific knowledge is irredeemably unverifiable and speculative, even if falsifiable, and (2) speculative ideas can be assessed *rationally*, by considering how well they solve the problems they were intended to solve, and the status of speculative philosophy changes dramatically. Abruptly, the activity of proposing and critically assessing possible speculative solutions to fundamental problems becomes a basic task of rational inquiry. Proposing speculative claims to knowledge, as possible solutions to fundamental problems of knowledge, becomes a vital, intellectually rigorous task for philosophy. It is not remotely prohibited by the valid Humean point that secure knowledge about the world cannot be acquired by pure thought alone. Intellectually respectable philosophical speculation must, of course, take note of science, and of other relevant developments in thought and life; the philosopher must try to be well-informed about everything, outside philosophy, relevant to the problems he attempts to solve.

aims-and-methods of diverse human endeavours — science, industry, government, the law, education, the arts — with the aim of contributing to the *improvement* of these endeavours, in particular by detecting and "curing" rationalistic neurosis. And despite the fact that "philosophy" means "the love of wisdom", academic philosophers have not yet cottoned on to the point that their proper, primary task at present is to get across to their fellow academics that a revolution is needed in the academic enterprise, so that the basic aim becomes to help humanity become wiser: see (Maxwell, 2003b).

3.10 Self-Preservation of Institutional Neurosis

The persistence of scientific and academic neurosis cannot be blamed entirely on the neurosis of philosophy. Institutional neurosis has its own built-in mechanisms for self-preservation. Thus, once standard empiricism is accepted, it becomes very difficult to challenge this philosophy of science within science itself. As we have seen, this is because standard empiricism, in insisting that ideas, in order to enter science, must be empirically testable, banishes (untestable) philosophy of science from science. Critical discussion of the adequacy, or inadequacy, of standard empiricism is simply excluded in toto from the arena where it is most needed, namely science itself. The failure of philosophers of science to justify standard empiricism is then interpreted by natural scientists as further justification for adopting the view that philosophy of science is best banished from science, an attitude that receives further support from the scientific sterility of neurotic philosophy of science. In this way neurosis tends to bring rationality into disrepute, thus discrediting the very intellectual tools needed to dismantle the neurosis.

And more generally, within the neurotic aim for academic inquiry of acquiring *knowledge*, one cannot easily raise questions about the desirability and rationality of this philosophy of inquiry, since to do so is to discuss rival views about what the aims and methods of inquiry ought to be, views that intermingle questions of value, fact, reason and possibility, and which therefore do not fit into inquiry devoted to the pursuit of knowledge.

And there are other, related, factors that tend to preserve neurotic institutional structures, once established. Reputations and careers of senior natural and social scientists have been built up on the basis of, and within, the neurotic structure of science: such powerful insiders will not take kindly to the suggestion that their reputations and careers have been based on mistakes.   Even those scientists who see through the neurotic structure of science will feel obliged to write, publish and teach in accordance with official, neurotic standards, simply in order to meet with acceptance and success.   Thus do public myths perpetuate themselves, even when individuals, responsible for perpetuating them, no longer believe in them.

3.11 The Neurosis of Psychoanalytic Theory and Practice

I began by reinterpreting neurosis as a *methodological* notion, a notion in the theory of rational aim-pursuing. The outcome was that it became not just meaningful, but true, to assert that science suffers from rationalistic neurosis.

This casts an interesting light on the intellectual standing of Freudian theory, and psychoanalytic theory more generally.   A number of philosophers and others have cast doubt on the intellectual standing of Freudian theory. Popper has criticized it for not being falsifiable, and hence not being scientific; Adolf Grünbaum has criticized it for having been either falsified, or not verified.[35]   And others have made other accusations.[36] But what the argument developed in this book has shown is that it is not Freud who fails to match up to the exacting standards of science; on the contrary, it is *science* that fails to match up to the exacting   intellectual   standards   of   Freudianism   reinterpreted methodologically. Science suffers from rationalistic neurosis, and needs methodological treatment.

This indicates the tremendous increase in scope and power that accrues from reinterpreting Freudianism (and psychoanalytic theory more generally) *methodologically*.   Thus reinterpreted, Freudianism becomes applicable to institutions, to anything that can be construed to pursue aims and to represent (and so misrepresent) aims being pursued.

One consequence of this methodological interpretation of Freud may

---

35. See (Popper, 1963, pp. 37-38); (Grünbaum, 1984).
36. See Grünbaum's book for references.

not, however, be so welcome to Freudians. This is that Freudianism itself is rationalistically neurotic! Freud saw himself as a scientist, contributing to the science of the psyche. In this respect, Freud was just another product of the bungled Enlightenment, which involves interpreting social inquiry as social *science*. But, in general, social inquiry ought to be developed as social methodology; and in particular, Freudianism ought to be developed as methodology. In so far as Freudians in particular, and psychoanalysts more generally, conceive of their disciple as science, or as seeking knowledge of the human psyche, their discipline suffers from rationalistic neurosis, and needs methodological therapy. Such a methodological reinterpretation of psychoanalysis — such a freeing of psychoanalysis from its rationalistic neurosis — would have profound implications for psychoanalytic theory and practice.

3.12 Conclusion

The natural and social sciences, and academic inquiry as a whole, suffer from rationalistic neurosis. This has far-flung damaging repercussions, both intellectual and social. At its most extreme, it means that humanity has not yet managed to develop traditions and institutions of learning well-designed from the standpoint of helping humanity learn how to create global civilization. This extraordinary state of affairs has persisted unrecognized in part because the intellectual failings are *philosophical* in character, and academic philosophy, which should have been actively correcting these persisting, institutional, philosophical blunders, has been obsessed with its own neurotic problems. We need an intellectual revolution to free the natural and social sciences, philosophy and academic inquiry as a whole from their damaging neuroses.

Chapter Four

What Is to Be Done?

4.1 Questions

Natural science, social science, philosophy, and the whole academic enterprise more generally, are locked into the methodological-institutional disease of rationalistic neurosis, with its built-in mechanisms of protection and self-preservation. The result is that humanity fails to have what it so urgently needs: traditions and institutions of inquiry rationally designed and devoted to helping humanity learn gradually to become wiser, more civilized.

At once a number of questions arise. What changes would have to be made to the academic enterprise if it were to free itself of neurosis? How can these changes be brought about given that academia has built-in institutional defence mechanisms which protect it against such changes being made? Would the powers that be, those who fund academia, primarily government and industry, permit the revolutionary changes to be made? Can we even be sure that these changes really are desirable, on intellectual and humanitarian grounds? Might not the outcome be worse than what we have at present? Are there not objections to carrying out the wholesale revolution in aims and methods being argued for here? Is it not absurd to suppose, for example, that *action* is intellectually more fundamental than *knowledge*, proposals for actions being discussed before relevant knowledge has been acquired?

And finally, even if the intellectual and institutional revolution being proposed here was brought about, what grounds do we have for supposing that this would have the slightest effect on what goes on in the big world beyond academia? The mere existence of intelligent thinking about what our problems are, and what we need to do about them, hardly suffices to ensure that it is this thinking that is influential in the context of political, industrial and commercial life. Indeed, academia today does already sustain such thinking, in Departments of Development, Policy Studies, Environmental Studies, Peace Studies and elsewhere; there is, however, little evidence that governments or the public pay much heed. How would bringing about the proposed revolution make a difference?

These are some of the questions I tackle in this chapter.

4.2 Recapitulation

But before tackling these questions, let me quickly review the argument that has been developed in the previous three chapters.

We have, in effect, two conceptions of inquiry before us: knowledge-inquiry and wisdom-inquiry. Both hold that a proper, fundamental, humanitarian aim of inquiry is to help promote human welfare by intellectual means. Knowledge-inquiry maintains that the best possible way to attain the humanitarian aim is for inquiry to pursue the sharply distinct intellectual aim of improving knowledge of factual truth. First, knowledge must be acquired; then, secondarily, it can be applied to help solve social problems, help promote human welfare. According to this view, only those considerations relevant to the acquisition of knowledge can be permitted to enter the intellectual domain of inquiry — academic publications, lectures and seminars — namely claims to factual knowledge, evidence, factual theories, problems of knowledge, and arguments, deductions and criticisms designed to help establish factual truth. Values, expressions of feelings and desires, political and moral considerations must all be excluded from the intellectual domain of inquiry, in order to ensure that authentic knowledge is acquired. If values, desires, feelings, political considerations were permitted to influence the intellectual domain of inquiry, academia would produce, not genuine, objective factual knowledge but fiction, propaganda and ideology, and the interests of humanity would be betrayed. Paradoxically, in order to help solve social problems, all expressions of human needs, desires and values must be ruthlessly excluded from academic inquiry, so that what is of genuine human use and value can be produced, namely value-neutral, factual knowledge.

At the core of knowledge-inquiry there is a philosophy of science, namely standard empiricism. This excludes not just expressions of feelings, desires and values but, even more extensively, everything that is not empirically *testable*.

Once upon a time academia, in Europe at least, was dominated by Christianity and Christian theology. With the decay of religion and the rise of modern science, the standard empiricist pursuit of knowledge came increasingly to dominate, especially with the institutional implementation of the Enlightenment idea of developing the social sciences alongside natural sciences.

But knowledge-inquiry, I have argued, is deeply neurotic. It is damagingly irrational in a wholesale, structural way. The standard empiricist view that the proper intellectual aim of science is to improve knowledge of factual truth, nothing being presupposed about the truth, is untenable. If put rigorously into scientific practice it would overwhelm science with endlessly many empirically successful disunified theories, and science would come to a standstill. Only the unrigorous implementation of standard empiricism makes scientific progress possible.

There are, to begin with, *metaphysical* assumptions implicit in the aim of science. Science has the aim of improving knowledge of *explanatory* truth, the truth being presupposed to be explanatory, or comprehensible. This aim is deeply problematic because the implicit metaphysical assumptions are bound to be more or less false. Problematic metaphysical assumptions implicit in the aim of science need to be made explicit, and arrayed in a hierarchy of metaphysical theses and associated methods, so that science has available a framework of fixed aims and methods within which more specific and problematic aims and methods can be improved with improving scientific knowledge.

Second, there are *value* assumptions implicit in the aims of science. Science does not just seek truth, or explanatory truth; more generally, inevitably and quite properly, it seeks *important* truth. But values associated with the aim of science are, if anything, even more problematic than metaphysical assumptions. Here, too, implicit value assumptions need to be made explicit, and arrayed in a hierarchy, so that problems associated with such value assumptions can be improved as science progresses.

Third, there are *political* assumptions implicit in the aims of science. Science is pursued in a social, economic, industrial and political context; in doing research and publishing results, scientists perform acts which can have social consequences, sometimes dramatic social consequences. The results of research are put to human use, as a result of industrial and commercial exploitation. Science is a part of implicit or explicit humanitarian, political, industrial and commercial programmes to bring about diverse changes in the social world. The problematic *political* assumptions implicit in the aims of science need to be made explicit, and arrayed in a hierarchy, so that these assumptions can be improved with improving knowledge.

Fourth, from its inception in the 17th century, science has been a part of an explicit humanitarian movement to better the lot of mankind, apparent, for example, in Francis Bacon's drive to create a kind of experimental or natural philosophy devoted to this end. This movement gained renewed impetus from the 18th century Enlightenment idea that we should learn from scientific progress how to make social progress towards an enlightened world. But in developing this idea, the philosophes of the Enlightenment made a series of blunders. In particular, they made the monumental blunder of seeking to develop social inquiry as social *science* rather than social *methodology* or social *philosophy*. They took it for granted that first, knowledge had to be acquired, and then, secondarily, applied to help solve social problems. This blunder was perpetuated throughout the 19th century and institutionalized in the first part of the 20th century with the creation of departments of social sciences in universities all over the world. But all this constitutes a massive blunder. The proper, *un-neurotic* way to develop the Enlightenment idea is to tackle social problems employing methods generalized from the progress-achieving methods of science. This involves at least employing problem-solving rationality (an improved version of Popper's critical rationalism). But ultimately it involves employing aim-oriented rationality, generalized from the actual progress-achieving methods of science, namely aim-oriented empiricism. Not just in science, but in life too, our fundamental aims are often deeply problematic. Above all, the aim of creating a better world is deeply problematic. Here, too, as in science, we need to array our problematic aims in a hierarchy, thus creating a framework of more or less fixed, unproblematic aims and methods, within which much more specific and problematic aims and methods can be improved as we live, as we act. Aim-oriented rationality needs to be built into all our other institutions, and not just into science; it needs to be built into our personal and social lives. Social inquiry has, as its fundamental intellectual task, to help us bring this about.

We have arrived at a new conception of inquiry: wisdom-inquiry. The intellectual and humanitarian aims of inquiry have become one and the same: to help promote wisdom in the real world by intellectual and educational means. Problems of living are intellectually fundamental;

problems of knowledge are subsidiary and secondary. Furthermore, reflecting this, social inquiry and the humanities are intellectually more fundamental than the natural sciences. The task of academic inquiry as a whole is not just to study the natural and social world; it has the task of learning from, teaching, communicating with and arguing with the social world. Academia is an intellectual and educational force for the promotion of personal and global wisdom.

Academia, as it exists at present in democratic nations, may perhaps be held to be a mixture of knowledge-inquiry and wisdom-inquiry. Overwhelmingly, however, knowledge-inquiry dominates. The fact that problems of living are discussed on the fringes of the academic enterprise, in departments of social policy, developmental and environmental studies, peace studies, international affairs, engineering, agriculture, medicine and elsewhere, does not suffice to establish that, to that extent, wisdom-inquiry is being implemented. Knowledge-inquiry includes discussion of how knowledge is to be *applied* to help solve social problems. What knowledge-inquiry, strictly speaking, cannot accommodate is the sustained intellectual activity of proposing and critically assessing possible solutions to problems of living — possible actions, policies, political programmes, philosophies of life, views about what is of value and how it is to be attained. This does go on in academia as it exists at present, but very much on the fringes, and not as the fundamental intellectual activity, as wisdom-inquiry demands.

Grounds for holding that knowledge-inquiry does dominate academia are given in chapter six of my book *From Knowledge to Wisdom*. There I looked at the following: (1) books about the modern university; (2) the philosophy and sociology of science; (3) statements of leading scientists; (4) Physics Abstracts; (5) Chemistry, Biology, Geo and Psychology Abstracts; (6) journal titles and contents; (7) books on economics, sociology and psychology; (8) philosophy. In 1984, the year *From Knowledge to Wisdom* was published, there can be no doubt whatsoever that knowledge-inquiry (or "the philosophy of knowledge" as I called it in the book) dominated academic inquiry.

Have things changed since then? Some things have moved in the direction of wisdom-inquiry. There is more awareness, today, of the importance of scrutinizing new scientific research projects, especially in biological and medical fields, for their moral implications. There is a greater recognition of the role of emotion in cognition.[1] Historians of science have placed great emphasis on the economic, social and political dimensions of science. Nevertheless, the revolution advocated by *From Knowledge to Wisdom*, and argued for here, has not occurred. There is still, amongst the vast majority of academics today, no awareness at all that a more intellectually rigorous and humanly valuable kind of inquiry than that which we have at present, exists as an option. In particular, social inquiry continues to be taught and pursued as social *science*, and not as social *methodology*. A year or two ago I undertook an examination, at random, of thirty-four introductory books on sociology, published between 1985 and 1997. Sociology, typically, is defined as "the scientific study of human society and social interactions" (Tischler, 1996, p. 4), as "the *systematic, sceptical study of human society*" (Macionis and Plummer, 1997, p. 4), or as having as its basic aim "to understand human societies and the forces that have made them what they are" (Lenski et al., 1995, p. 5). Some books take issue with the idea that sociology is the *scientific* study of society, or protest at the male dominated nature of sociology (see, for example, Abott and Wallace, 1990, p. 3 and p. 1). Nowhere did I find a hint of the idea that a primary task of sociology, or of social inquiry more generally, might be to help build into the fabric of social life progress-achieving methods, generalized from those of science, designed to help humanity resolve its conflicts and problems of living in more cooperatively rational ways than at present.

4.3 From Knowledge to Wisdom

What needs to be changed, if present-day knowledge-inquiry is to be converted into wisdom-inquiry?

The changes I now indicate are not arbitrary: they all arise from the simple demand that academic inquiry should satisfy elementary requirements for *rigour*; they all arise, that is, from the simple demand

---

1. The fundamental role of emotion in cognition, as far as neuroscience is concerned, has been stressed especially by Damasio: see (Damasio, 1994, 2000). Nearly two decades earlier, I stressed that emotion is essential to rationality, to rational inquiry, and to science: see (Maxwell, 1976a, pp. 3-13, 64-69, 81-88, 111-119, 139-157, 165-170, 184-230). The point is stressed again in (Maxwell, 1984a: see pp. 6-8, 78-79, 181-189, 251-253 and 264-267).

that inquiry should put problem-solving and aim-oriented rationality into practice in seeking to help humanity become wiser.

(1) There needs to be a change in the basic intellectual *aim* of inquiry, from the growth of knowledge to the growth of wisdom — wisdom being taken to be the capacity to realize what is of value in life, for oneself and others, and thus including knowledge, understanding and technological know-how.

(2) There needs to be a change in the nature of academic *problems*, so that problems of living are included, as well as problems of knowledge. Furthermore, problems of living need to be treated as intellectually more fundamental than problems of knowledge.

(3) There needs to be a change in the nature of academic *ideas*, so that proposals for action are included as well as claims to knowledge. Furthermore, proposals for action need to be treated as intellectually more fundamental than claims to knowledge.

(4) There needs to be a change in what constitutes intellectual *progress*, so that progress-in-ideas-relevant-to-achieving-a-more-civilized-world is included as well as progress in knowledge, the former being indeed intellectually fundamental.

(5) There needs to be a change in the idea as to where inquiry, at its most fundamental, is located. It is not esoteric theoretical physics, but rather the thinking we engage in as we seek to achieve what is of value in life.

(6) There needs to be a dramatic change in the nature of social inquiry (reflecting points 1 to 5). Economics, politics, sociology, and so on, are not, fundamentally, *sciences*, and do not, fundamentally, have the task of improving knowledge about social phenomena. Instead, their task is threefold. First, it is to articulate problems of living, and propose and critically assess possible solutions, possible actions or policies, from the standpoint of their capacity, if implemented, to promote wiser ways of living. Second, it is to promote such cooperatively rational tackling of problems of living throughout the social world. And third, at a more basic and long-term level, it is to help build the hierarchical structure of aims and methods of aim-oriented rationality into personal, institutional and global life, thus creating frameworks within which progressive

improvement of personal and social life aims-and-methods becomes possible. These three tasks are undertaken in order to promote cooperative tackling of problems of living — but also in order to enhance empathic or "personalistic" understanding between people as something of value in its own right. Acquiring knowledge of social phenomena is a subordinate activity, engaged in to facilitate the above three fundamental pursuits.

(7) Natural science needs to change, so that it includes at least three levels of discussion: evidence, theory, and research aims. Discussion of aims needs to bring together scientific, metaphysical and evaluative consideration in an attempt to discover the most desirable and realizable research aims.

(8) There needs to be a dramatic change in the relationship between social inquiry and natural science, so that social inquiry becomes intellectually more fundamental from the standpoint of tackling problems of living, promoting wisdom.

(9) The way in which academic inquiry as a whole is related to the rest of the human world needs to change dramatically. Instead of being intellectually dissociated from the rest of society, academic inquiry needs to be communicating with, learning from, teaching and arguing with the rest of society — in such a way as to promote cooperative rationality and social wisdom. Academia needs to have just sufficient power to retain its independence from the pressures of government, industry, the military, and public opinion, but no more. Academia becomes a kind of civil service for the public, doing openly and independently what actual civil services are supposed to do in secret for governments.

(10) There needs to be a change in the role that political and religious ideas, works of art, expressions of feelings, desires and values have within rational inquiry. Instead of being excluded, they need to be explicitly included and critically assessed, as possible indications and revelations of what is of value, and as unmasking of fraudulent values in satire and parody, vital ingredients of wisdom.

(11) There need to be changes in education so that, for example, seminars devoted to the cooperative, imaginative and critical discussion of problems of living are at the heart of all education from five-year-olds onwards. Politics, which cannot be taught by knowledge-inquiry, becomes central to wisdom-inquiry, political

creeds and actions being subjected to imaginative and critical scrutiny.

(12) There need to be changes in the aims, priorities and character of pure science and scholarship, so that it is the curiosity, the seeing and searching, the knowing and understanding of individual persons that ultimately matters, the more impersonal, esoteric, purely intellectual aspects of science and scholarship being means to this end. Social inquiry needs to give intellectual priority to helping empathic understanding between people to flourish (as indicated in 6 above).

(13) There need to be changes in the way mathematics is understood, pursued and taught. Mathematics is not a branch of knowledge at all. Rather, it is concerned to explore problematic *possibilities*, and to develop, systematize and unify problem-solving methods.

(14) Literature needs to be put close to the heart of rational inquiry, in that it explores imaginatively our most profound problems of living and aids personalistic understanding in life by enhancing our ability to enter imaginatively into the problems and lives of others.

(15) Philosophy needs to change so that it ceases to be just another specialized discipline and becomes instead that aspect of inquiry as a whole that is concerned with our most general and fundamental problems — those problems that cut across all disciplinary boundaries. Philosophy needs to become again what it was for Socrates: the attempt to devote reason to the growth of wisdom in life.[2]

This is the revolution we need to bring about in our traditions and institutions of learning, if they are to be properly and rationally designed to help us learn how to make progress towards a wiser world.

4.4 Objections

I now consider a variety of objections that may be raised against the

---

2. For further discussion of the revolution in academic aims and methods that I am advocating here, see my *From Knowledge to Wisdom*.

intellectual and institutional revolution for which I have been arguing so far.

It may be objected that it is all to the good that the academic enterprise today does give priority to the pursuit of knowledge over the task of promoting wisdom and civilization. Before problems of living can be tackled rationally, knowledge must first be acquired. This is the objection which, I imagine, most academics will want to make.

I have six replies to this objection:

First, even if the objection were valid, it would still be vital for a kind of inquiry designed to help us build a better world to include rational exploration of problems of living, and to ensure that this guides priorities of scientific research (and is guided by the results of such research).

Second, the validity of the objection becomes dubious when we take into account the considerable success people met with in solving problems of living in a state of extreme ignorance, before the advent of science. We still today often arrive at solutions to problems of living in ignorance of relevant facts.

Third, the objection is not valid. In order to articulate problems of living and explore imaginatively and critically possible solutions (in accordance with problem-solving rationality) we need to be able to act in the world, imagine possible actions and share our imaginings with others: in so far as some common sense knowledge is implicit in all this, such knowledge is required to tackle rationally and successfully problems of living. But this does not mean that we must give intellectual priority to acquiring new relevant knowledge before we can be in a position to tackle rationally our problems of living.

Fourth, simply in order to have some idea of what kind of knowledge or know-how it is *relevant* for us to try to acquire, we must *first* have some provisional ideas as to what our problem of living is and what we might do to solve it. Articulating our problem of living and proposing and critically assessing possible solutions needs to be intellectually prior to acquiring relevant knowledge simply for this reason: we cannot know what new knowledge it is *relevant* for us to acquire until we have at least a preliminary idea as to what our problem of living is, and what we propose to do about it. A slight change in the way we construe our problem may lead to a drastic change in the kind of knowledge it is relevant to acquire. Thus, for example, changing the way we construe problems of health, to include *prevention* of disease (and not just curing of disease) leads to a dramatic change in the kind of knowledge we need to acquire (importance of exercise, diet etc.). Again, including the importance of avoiding *pollution* in the problem of creating wealth by means of industrial development leads to the need to develop entirely new kinds of knowledge.

Fifth, relevant knowledge is often hard to acquire; it would be a disaster if we suspended life until it had been acquired. Knowledge of how our brains work is presumably highly relevant to all that we do but clearly, suspending rational tackling of problems of living until this relevant knowledge has been acquired would not be a sensible step to take. It would, in any case, make it impossible for us to acquire the relevant knowledge (since this requires scientists to act in doing research). Scientific research is itself a kind of action carried on in a state of relative ignorance.

Sixth, the capacity to act, to live, more or less successfully in the world, is more fundamental than (propositional) knowledge. Put in Rylean terms, "knowing how" is more fundamental than "knowing that" (Ryle, 1949, ch. II). All our knowledge is but a development of our capacity to act. Dissociated from life, from action, knowledge stored in libraries is just paper and ink, devoid of meaning.[3] In this sense, problems of living are more fundamental than problems of knowledge (which are but an aspect of problems of living); giving intellectual priority to problems of living quite properly reflects this point. For a development of this point, see (Maxwell, 1984a, pp. 174-181); see also (Maxwell, 2001c, especially chs. 5-7 and 9).

As we saw in the last chapter, a kind of inquiry that gives priority to

---

3. Goal-pursuing things can be put into three categories: (a) those that lack all sentience and consciousness (robots); (b) those that are sentient but not conscious (some animals); (c) and those that are sentient and conscious (persons). "Belief" and "knowledge", likewise, have three different meanings depending on whether the associated goal-pursing things are in category (a), (b) or (c), these meanings becoming enriched as one moves from (a) to (b) to (c). (See Maxwell, 2001c, ch. 7 for a discussion.) The crucial point, however, is that, dissociated from all goal-pursuing things, whether of type (a), (b) or (c), belief and knowledge do not exist. Sounds in the air, marks on paper or records in computers only constitute expressions or formulations of belief or knowledge in so far as they are interpretable as such with respect to goal-pursuing things of type (a), (b) or (c).

tackling problems of knowledge over problems of living violates the most elementary requirements of rationality conceivable. If the basic task is to help humanity create a better world, then the problems that need to be solved are, primarily, problems of living, problems of action, not problems of knowledge. This means that to comply, merely, with Popper's conception of critical rationalism or, better, problem-solving rationality, discussed in chapter three, the basic intellectual tasks need to be (1) to articulate problems of living, and (2) to propose and critically assess possible solutions, possible more or less cooperative human *actions*. (1) and (2) are excluded, or marginalized, by a kind of inquiry that gives priority to the task of solving problems of knowledge. And the result will be a kind of inquiry that fails to create a reservoir of imaginative and critically examined ideas for the resolution of problems of living, and instead develops knowledge often unrelated to, or even harmful to, our most basic human needs.

It may be objected that in employing aim-oriented rationality in an attempt to help create a more civilized world, in the way that has been indicated, wisdom-inquiry seeks to change the whole of the social world, and thus falls foul of Popper's strictures against Utopian social engineering: see (Popper, 1969, vol. 1, ch. 9; 1974, pp. 64-92).

But wisdom-inquiry is not remotely like Popper's "Utopian social engineering". Popper's "Utopian social engineer" sets out to seize power and impose an overall plan, an ideology, on society by force. Wisdom-inquiry seeks to promote more cooperatively rational resolving of social conflicts and problems by means of imaginative and critical exploration of possibilities, debate, argument, learning from past successes and failures, the active engagement in and promotion of rationally cooperative discussion with the public and all those concerned. In so far as piecemeal social engineering, of the kind advocated by Popper, is indeed the rational way to make progress towards a more civilized world, this will be advocated by wisdom-inquiry and the New Enlightenment.

There are, however, reasons for holding that piecemeal social engineering is inadequate. When we take into account the unprecedented *global* nature of many of our most serious problems, (the outcome of solving the first great problem of learning but failing to solve the second), we may well doubt that piecemeal social engineering is sufficient. Popper's distinction between Utopian and piecemeal social engineering is in fact much too crude: it overlooks a third strategy, that of wisdom-inquiry, aim-oriented rationalistic social engineering, as it might be called, with its emphasis on developing increasingly cooperatively rational resolutions of human conflicts and problems in full recognition of the inherently problematic nature of the aim of achieving greater civilization.

Finally, it is important to differentiate between the aim of changing (1) the whole of society, and (2) some specific feature or aspect of society which, as it changes, will have repercussions for the whole of society. (1) is perhaps "holistic" in a sense in which Popper correctly condemns, but (2) is not. Examples of type-(2) campaigns are feminism, and the green movement. Both seek to change a specific aspect of the social world: the inequality of women and men, and the way the modern world damages the environment. But both have repercussions for the whole of society. Wisdom-inquiry (which would include aspects of feminism and the green movement) is a type-(2) endeavour: it seeks to promote the quite specific change of an increase in cooperatively rational tackling of problems of living in the social world, but this change would have repercussions for the whole of society. (For further discussion see Maxwell, 1984a, pp. 189-198.)

It might of course be held to be Utopian to expect that academia could change in the wholesale way required if wisdom-inquiry is to replace knowledge-inquiry. But again, such a wholesale change may affect every branch and aspect of academic inquiry, but is, nevertheless, a change of a quite specific kind: the repeated acknowledgement of problematic aims. And besides, the revolution in aims and methods involved in moving from knowledge-inquiry to wisdom-inquiry would, inevitably, be done bit by bit, in a gradual, piecemeal fashion.

All those to any degree influenced by Romanticism and the counter-Enlightenment will object strongly to the idea that we should learn from scientific progress how to achieve social progress towards civilization; they will object strongly to the idea of allowing conceptions of rationality, stemming from science, to dominate in this way, and will object even more strongly to the idea, inherent in the new Enlightenment, that we need to create a more aim-oriented rationalistic social world.[4]

Directed at knowledge-inquiry and the traditional Enlightenment, objections of this kind have some validity; but directed at wisdom-inquiry and the new Enlightenment, they have none. As I pointed out in the last chapter, and as I have emphasized elsewhere,[5] aim-oriented rationality amounts to a synthesis of traditional Rationalist and Romantic ideals, and not to the triumph of the first over the second. In giving priority to the realization of what is of value in life, and in emphasizing that rationality demands that we seek to improve aims as we proceed, the new Enlightenment requires that rationality integrates traditional Rationalist and Romantic values and ideals of integrity. Imagination, emotion, desire, art, empathic understanding of people and culture, the imaginative exploration of aims and ideals, which tend to be repudiated as irrational by traditional Rationalism, but which are prized by Romanticism, are all essential ingredients of aim-oriented rationality. Far from crushing freedom, spontaneity, creativity and diversity, aim-oriented rationality is essential for the desirable flourishing of these things in life. But aim-oriented rationality and wisdom-inquiry also, correctly, value and use improved versions of Rationalist ideals: reason, science, observation and experiment, the search for truth, objectivity, respect for fact and valid argument. The idea that Rationalist and Romantic ideals are in collision, science and reason crushing spontaneity, imagination, freedom, art and empathic understanding, is all the result of the failure of Romanticism to correct the failings of traditional Rationalism and the Enlightenment. If the Enlightenment had been developed free of its blunders, Romanticism would have emerged and flourished as a continuation and enrichment of the Enlightenment, as a part of (aim-oriented) rationality, and not opposed to reason. Traditional Rationalist endeavours, such as science, mathematics, rational argument, the attempt to resolve conflicts rationally, all need Romantic ideals and virtues if they are to succeed: imagination, empathic understanding, inspiration, spontaneity, individuality. And likewise, traditional Romantic endeavours, such as the arts, friendship

---

4. For literature protesting against the influence of scientific rationality in various contexts and ways, see for example: (Berlin, 1999; Laing, 1965; Marcuse, 1964; Roszak, 1973; Berman, 1981; Schwartz ,1987; Feyerabend, 1978 and 1987; Appleyard, 1992).

5. (Maxwell, 1984a, pp. 63-64, pp. 85-91 and pp. 117-118; 1976a, especially chs. 1 and 8-10; 2001c, ch. 5 and 9).

and love, need Rationalistic virtues to succeed: a concern for fact and truth, objectivity, freedom from dogma, intellectual independence and integrity. Both Rationalist and Romantic endeavours degenerate if split off from each other, and gagged and bound in rationalistic neurosis. As I have said, we need to put the heart and the head in touch with each other, so that we may develop heartfelt minds and mindful hearts.

Many historians and sociologists of science deny that there is any such thing as scientific method or scientific progress, and will thus find the basic idea of this book absurd.[6] These writers are encouraged in their views by the long-standing failure of scientists and philosophers of science to explain clearly what scientific method is, and how it is to be justified. This excuse for not taking scientific method and progress seriously is, however, no longer viable: as we have seen, reject standard empiricism in all its forms, and it becomes clear how scientific method and progress are to be characterized and justified, in a way which emphasizes the rational interplay between evolving knowledge and evolving aims and methods of science.[7] In a world dominated by the products of scientific progress it is quixotic in the extreme to deny that such progress has taken place.

Finally, those of a more Rationalist persuasion may object that science is too different from political life for there to be anything worthwhile to be learnt from scientific success about how to achieve social progress towards civilization.[8] (a) In science there is a decisive procedure for eliminating ideas, namely, empirical refutation: nothing comparable obtains, or can obtain, in the political domain. (b) In science experiments or trials may be carried out relatively painlessly (except, perhaps, when new drugs are being given in live trials); in life, social experiments, in that they involve people, may cause much pain if they go wrong, and may be difficult to stop once started. (c) Scientific progress requires a number of highly intelligent and motivated people to pursue

6. See (Bloor, 1976; Barnes and Bloor, 1981; Latour, 1987; Feyerabend, 1978 and 1987). These authors might protest that they do not deny scientific knowledge, method, progress or rationality as such, but deny, merely, that the sociology of knowledge can legitimately appeal to such things, or deny extravagant claims made on behalf of these things. See, however, the sparkling criticism by (Sokal and Bricmont, 1998, ch. 4).

7. See chapter one and the appendix; and see (Maxwell, 1998, especially chs. 1-6).

8. Nicholas Rescher (personal communication); (Durant, 1997).

science on behalf of the rest of us, funded by government and industry; social progress requires almost *everyone* to take part, including the stupid, the criminal, the mad or otherwise handicapped, the ill, the highly unmotivated; and in general there is no payment. (d) Scientists, at a certain level, have an agreed, common objective: to improve knowledge. In life, people often have quite different or conflicting goals, and there is no general agreement as to what civilization ought to mean, or even whether it is desirable to pursue civilization in *any* sense. (e) Science is about fact, politics about value, the quality of life. This difference ensures that science has nothing to teach political action (for civilization). (f) Science is male-dominated, fiercely competitive, and at times terrifyingly impersonal (Harding, 1986); this means it is quite unfit to provide any kind of guide for life.

Here, briefly, are my replies. (a) Some proposals for action can be shown to be unacceptable quite decisively as a result of experience acquired through attempting to put the proposal into action. Where this is not possible, it may still be possible to assess the merits of the proposal to some extent by means of experience. If assessing proposals for action by means of experience is much more indecisive than assessing scientific theories by means of experiment, then we need, all the more, to devote our care and attention to the former case. (b) Precisely because experimentation in life is so much more difficult than in science, it is vital that in life we endeavour to learn as much as possible from (i) experiments that we perform in our imagination, and (ii) experiments that occur as a result of what actually happens. (c) Because humanity does not have the aptitude or desire for wisdom that scientists have for knowledge, it is unreasonable to suppose that progress towards global wisdom could be as explosively rapid as progress in science. Nevertheless progress in wisdom might go better than it does at present. (d) Cooperative rationality is only feasible when there is the common desire of those involved to resolve conflicts in a cooperatively rational way. (e) Aim-oriented rationality can help us improve our decisions about what is desirable or of value, even if it cannot reach decisions for us. (f) In taking science as a guide for life, it is the progress-achieving methodology of science to which we need to attend. It is this that we need to generalize in such a way that it becomes fruitfully applicable, potentially, to all that we do. That modern science is male-dominated, fiercely competitive, and at times terrifyingly impersonal should not deter us from seeing what can be learned from the progress-achieving methods of science — unless, perhaps, it should turn out that being male-dominated, fiercely competitive and impersonal is essential to scientific method and progress. (But this, I submit, is not the case.)

So far I have considered objections to the desirability of wisdom-inquiry; but what about its feasibility? Would not the defence mechanisms of neurosis, built into the institutional structure of academic inquiry, prevent the required revolution in aims and methods from taking place? And if these neurotic defences could somehow be overcome, what assurance can we have that anything worthwhile would emerge from social inquiry taking up its wisdom-inquiry task of articulating our problems of living and proposing and critically assessing possible solutions? Might not this degenerate into no more than standard political, religious and ideological camps advocating their standard prescriptions? Are not academics, sheltered in their ivory towers from abrasive reality, the last people to come up with new, good, practical ideas as to how we can solve our problems of living? Is there not a catch 22 here? Do not academics first need wisdom before they can promote wisdom wisely? If the public are to be invited to join in the debate, how can academia sift out what is genuinely fruitful from the banal, the crankish, the mad?

Here are my replies to these objections. The defence mechanisms of neurosis do indeed make it difficult to get wisdom-inquiry up and running; but this obstacle is not insurmountable. What is required, intellectually, to defeat neurosis is repeated application of the principle: assumptions that are substantial, influential, problematic and implicit need to be made explicit, so that they can be critically assessed, so that alternatives can be considered, in an attempt to improve such assumptions. What is required, institutionally and socially, is a large enough group of scientists and academics who understand the urgent need to bring wisdom-inquiry into existence, and who are prepared to work, in their professional capacities, to help bring the required changes about. As soon as academics begin to believe that the revolution will occur, sooner or later, then they will want to be in on the act. Pioneers will be rewarded, once the revolution has occurred!

There can, of course, be no guarantee that worthwhile intellectual work would emerge when or if the revolution occurs. A major task confronting those involved in bringing the revolution about would be to produce work that is intelligent, imaginative, humane, relevant and practical. In academia as it is today, much goes on that is fraudulent: esoteric, jargon-ridden work in cultural studies, sociology, philosophy, and in other areas of social science and the humanities. Granted the profoundly neurotic state of the status quo, this is almost to be expected. Freeing academia of neurosis ought to decrease the volume of fraudulent academic activity. Suddenly, social inquiry and the humanities become intellectually fundamental; they assume responsibility for helping humanity resolve its conflicts and problems of living in increasingly cooperatively rational ways. Given this profoundly serious task, the temptation to indulge in anti-rationalist posturing and scholastic drivel must surely diminish. Once anti-rationalists appreciate that what they are really opposed to is not reason, but a characteristic kind of *irrationality* masquerading as reason, the indispensability of reason may come to be more widely appreciated. It is of course vital that academics learn to treat non-academics with things to say as colleagues, and not merely as objects of study. A major task for academics must be to discover and emphasize good ideas, wherever they come from. As for the danger that academics split up into standard opposing political and ideological camps, aim-oriented rationality is designed to prevent this from happening, in that it leads opponents to get clear about what is common ground, what disputed. It is a methodology for the progressive resolution of such unfruitful clashes. But it does not, of course, guarantee success; there is no guarantee that disputants will want to employ it, or will use it honestly. If you seek to acquire wisdom it may well help to be wise to begin with, but the same could be said about acquiring knowledge. None of these objections constitute grounds for continuing with academia as it is at present, crippled by rationalistic neurosis.

4.5 What Are Our Most Serious Problems of Living and How Could Wisdom-Inquiry Help?

Strictly speaking all living is problem-solving.[9] Just how inherently complex and challenging are the problems associated with even the

---

9. This point was made in the last chapter, note 21. See also (Maxwell, 1984a, p. 92).

simplest of human actions, such as recognizing a mug of tea, picking it up and taking a sip, is revealed when artificial intelligence experts attempt to create a robot able to imitate such an action.

Nevertheless, much of what we do is done instinctively, spontaneously, our wonderfully educated brains solving the associated intractable problems for us without any awareness of effort or difficulty on our part. Open our eyes, and a rich scene is instantly revealed to us, the result of an amazingly complex mass of problems being solved almost instantaneously by our brains. Furthermore, it is not just "commonplace" actions that we may perform spontaneously, without any conscious effort at problem-solving: actions of great value, involving artistic creation, friendship, love, care for others, may be done spontaneously, without any sense of engaging in difficult problem-solving. The ideal — even the *rational* ideal — is to live, and to realize what is of value in the circumstances of our life — instinctively and spontaneously. Rationality requires that we only stop to engage in conscious, explicit problem-solving when we get into difficulties, when what we feel tempted to do instinctively would not have desirable outcomes, or when we do not know what to do. It may be, of course, that we have problems that we do not recognize, either because they are misconstrued, or because we fail to appreciate that desirable but problematic ends are available.

What our problems are vary, of course, from time to time, place to place, from person to person, and from society to society. Most of us, at some stage in our lives, face problems of friendship and love, of employment, health, parental care, discovering that which can make our lives genuinely worthwhile. Nevertheless, a nomad struggling to keep going on the edge of the Sahara faces life problems very different from those facing a young man with wealthy parents in Europe or the USA.

What are our current most urgent global problems – problems that affect most of us, or many of us? Here is a possible list of our nine most serious:

(1) Third world poverty. The vast, unjust inequalities of wealth across the globe.
(2) War and the threat of war, including terrorism.
(3) Environmental problems: destruction of tropical rain forests and

other natural habitats, rapid extinction of species, depletion of vital natural resources such as oil, pollution of sea, earth and air, global warming.

(4) Population growth.

(5) Dictatorial regimes.

(6) Violation of basic civil rights.

(7) Annihilation of languages, cultures, traditional ways of life.

(8) Threat posed by conventional, chemical, biological and nuclear armaments.

(9) AIDS epidemic.

No doubt a somewhat different choice could be made of our most urgent global problems. Some would place far greater emphasis on the evils of globalization and exploitative development; others on the war against modern terrorism. But let us, for the sake of the argument, take these to be our most serious problems.

One immediate reaction might be that most of the above are not really problems in the conventional sense at all, in that we know exactly what needs to be done progressively to tackle them. They persist because those who have the power to do something about them choose not so to act. Taking each in turn, what needs to be done in response to these problems is as follows.

(1*) An end to third world debt. A massive effort by the wealthy first world to help the poor of the third world to help themselves. Provision of education, training, resources, markets. An end to unnecessary infant mortality, child labour and slavery, lack of adequate housing, food and water, provision of health care to all. A more equable distribution of land ownership, especially in south America, south Africa, Indonesia and elsewhere. And, more generally, a more just distribution of the world's wealth, and fairer trade arrangements.

(2*) Strengthening of the UN as global peacekeeper; sustained pressure to make the UN more democratic; battle against terrorism pursued as an international police operation against crime, not as a war.

(3*) Aid to countries containing tropical rain forests to assist (a) development which does not destroy forests, and (b) policing of criminal destruction of forest; international measures to bring extinction of species to an end; development of alternative sources of energy to oil; development of sustainable, non-polluting world industry and agriculture.

(4*) More resources for making available contraceptives to the poorest; rapid increase in the standard of living of the world's poorest (which tends to lead to decrease in population growth).

(5*) Dictators to be treated as hostage-takers by the democratic world community. Negotiation, pressure, inducements to abandon power, aid to the opposition, but the lives and welfare of the hostages — the vast majority of those living under the dictatorship — being the primary consideration.

(6*) An end to the violation of basic civil rights; penalties for those governments that do violate such rights.

(7*) Active preservation of languages, cultures, traditional ways of life.

(8*) End to the arms trade; control and reduction of conventional arms; world-wide banning of chemical, biological and nuclear armaments (including present nuclear powers, of course, such as the USA and UK), and a system of world-wide inspection.

(9*) Drugs made available to the poor for HIV and AIDS; world-wide educational campaign about how not to contract HIV; continuing research into discovery of vaccination against HIV.

In order to implement these policies internationally it will be necessary to have a democratic world government determined to make them a priority. We need, then, a tenth policy:

(10*) Creation of a democratic world government.

All this, it may be objected, ought to be obvious to any reasonably well-informed and humane person. It does not need an academic industry to establish these elementary points. It is not ignorance that prevents us from pushing through the above ten policies, but lack of will. The poor of the third world lack the means to do anything about the matter. Those of us fortunate enough to live in the wealthy first world do not care enough about the matter to devote ourselves to political activity on behalf of these policies; or we actually do not want the policies adopted, fearing that they would have the consequence that our way of life would suffer if they were.

I have two points to make in reply to this objection.

First, it is not clear that (1*) to (10*) do specify policies which, if implemented, would solve (1) to (10). In some cases the "solution" is little better than the statement that the problem is solved. To say "Active preservation of languages, cultures, traditional ways of life" scarcely tells us what is to be done, and how this will achieve the desired end. Nothing is said about the decay of cultures and traditional ways of life, not as a result of exploitation and violence from without (which might, perhaps, be prevented), but as a result of the remorseless, irresistible processes that go with being in contact with global culture and technology, such as TV and tourism. Nothing is said about how to overcome the problem that many cultures and traditional ways of life need to change because of inherent injustice and dogmatism, such as that associated with treatment of women, or with religious fundamentalism. The problem, here, is to discover how to bring about much needed change without destruction of an entire culture and tradition. Even if (some of) (1*) to (10*) do indicate policies which, if implemented, might solve (some of) (1) to (10), almost everything would depend on how, locally and in detail, these policies are put into practice. In connection with (1*), for example, there are endless cases of aid having adverse effects: aid is siphoned off to make the wealthy wealthier; it promotes corruption; it destroys local industry and agriculture; it destroys self-reliance and makes recipients dependent on aid, almost like a drug. Almost all of (1*) to (10*) could be implemented in ways which would cause more harm than good, even if implemented with the best of intentions. If governments, or world government, take(s) charge of implementing (1*) to (9*), there is the standard danger involved in centralizing power, giving power to government. The claim that it is entirely obvious what needs to be done to solve problems (1) to (9) is unfounded. It is rather like claiming, within the context of science, that physicalism solves the scientific problem of the ultimate nature of the universe. One should regard (1*) to (10*) as constituting a kind of overall "blueprint" for policy which, at best, requires much further intelligent thought and experience — aim-oriented rationalistic research allied to action and aware of past successes and failures — if it is to be implemented successfully everywhere in particular, local circumstances.

Second, even if we imagine that (1*) to (10*) have been developed in detail, as a result of experience and research, of trial and error (as Popper would say), so as to solve (1) to (9) if enacted, this leaves unaffected a gigantic task of education. It is reasonable to assume that, at present, only a small percentage of those living in first world countries would favour putting (1*) to (10*) into practice. For (1*) to (10*) to become priorities for policy for democratic governments in north America, Europe, Japan, Australia and elsewhere, these priorities must be popular. A majority of the relevant electorates must believe that the policies are necessary, desirable, feasible. And this, in turn, requires education. Knowledge-inquiry is ill-equipped to provide the required kind of political, environmental, economic and moral education. Granted neurotic, knowledge-inquiry intellectual standards, education concerning the problems, (1) to (9), and the proposed solutions, (1*) to (10*), could only amount to some kind of sustained attempt at indoctrination, which would be bound to fail. Granted un-neurotic wisdom-inquiry, the educational task would be to encourage sustained exploration and discussion of how the problems (1) to (9) are to be progressively resolved. The fundamental task of the academic world would be to promote such exploration and discussion in the media, in government, in public life in general, as well as in schools and universities.

Let us suppose that (1) to (9) do indeed encapsulate our most urgent current global problems, (1*) to (10*) being the policies we need to adopt if we are progressively to resolve these problems. Let us suppose, further, that the intellectual revolution has occurred, academia puts wisdom-inquiry into practice, and actively and intelligently develops and argues for the above policies. What reason do we have to suppose that this would affect, in any way whatsoever, what goes on in the big world beyond academia?

There is, after all, nothing novel about any of these policies. Some have considerable popular support, some are actively supported by campaigns or charities such as the green movement or amnesty international, and some have been adopted, officially at least, by some governments: all have been argued for at some time or other. Some academics have concerned themselves, in their professional capacity, with aspects of some of these policies, within the current framework of knowledge-inquiry. Despite this, at present the world attempts to put these policies into practice in only a very feeble or hypocritical fashion, if at all. The above policies are not exactly at the centre of attention in the media, and in current political debate. And of course many people, groups, corporations, institutions and countries oppose many of these policies, and pursue courses of action which are the very opposite of what these policies require (most notably, at the time of writing, the USA). How would this situation be changed by the advent of wisdom-inquiry?

It is of course conceivable that wisdom-inquiry would have no impact. Having in existence traditions and institutions of learning devoted to helping us discover wisdom and civilization in a genuinely rational, un-neurotic fashion may well be necessary for us to make progress towards a wise and civilized world, but it cannot be *sufficient*. Academia might dream up magnificent policies, practical and effective, which are just ignored by the rest of the world.

There are grounds, however, for holding that this would be unlikely to happen. Wisdom-inquiry, taken up by the academic establishment, would transform the *status quo*. Such an academic establishment would pour forth a flood of books, articles, television programmes, newspaper articles, lectures, interviews and reports emphasizing and re-emphasizing the importance of taking the above policies seriously, illustrating the disasters that ensue if this is not done, and tackling the multitude of problems that need to be solved if these policies are to be implemented. The outcome of all this would be to keep the problems of civilization constantly before the public mind. In the public arena, there would be persistent discussion of these problems, and what to do about them. This would affect radically what people believe to be possible. It would influence political parties in their choice of policies and priorities; and, ultimately, it would affect government action. The industrial and commercial world would need to take into account the interests, the concerns of consumers who take seriously the possibility of creating a more civilized world.

Academics have a multitude of contacts with government, the civil service, industry, the financial world, the media, the judiciary, local government, education. If all these contacts are used to argue the case for the adoption and implementation of the above policies, and the arguments and policies are sound, this is bound to influence attitudes towards their acceptability.

Wisdom based academic inquiry does not just inform the rest of the population about the results of its deliberations. It also does all it can to promote, to provoke, cooperative rational problem-solving in personal and social life. And, unlike knowledge based inquiry, it eagerly seeks to learn from non-academics what approaches to problems, what policies and strategies have met with success and failure. Wisdom-inquiry strives to make itself available as a resource for cooperative rational problem-solving in life in a way in which knowledge-inquiry does not.

Wisdom-inquiry would also influence the social world via education. Students who do three years of wisdom-inquiry (as opposed to knowledge-inquiry) in whatever field — economics, natural science, sociology, philosophy or literature — are bound to see the world and its problems differently. The idea that humanity can learn how to solve its problems in increasingly rationally cooperative ways, thus making progress towards global civilization, will be familiar and obvious, and not utopian nonsense. Graduates will be aware of more or less specific changes that could be made to current institutions, practices, habits of thought, which would promote social wisdom and civilization. Policies at present impossible, because only a few people see the need for them, or are able to imagine them as practical, become possible because a sufficient number of graduates are in positions of influence or power who know full well that these policies have been subjected to sustained critical scrutiny, and have emerged as viable.

It is of course true that, in industrially advanced societies only a certain percentage of the population receives a university education. But wisdom-inquiry would have an impact far beyond that minority. In the first place, as I have indicated, graduates tend to occupy positions of influence or power in society. Secondly, many teachers in primary and secondary schools are graduates; and all teachers, in attending a teacher's training course, would be exposed to wisdom-inquiry. Thirdly, what is taught in primary and secondary schools and their equivalent is massively influenced by what goes on at university level. Wisdom-inquiry would be located, not just in universities, but in colleges of further education, in secondary and primary schools as well. Everyone would receive wisdom education.

In a democracy, people get the governments that they deserve. A politically naive electorate may well be bamboozled by charismatic and corrupt politicians into electing governments that pursue policies of disaster, as in what used to be Yugoslavia, to take an especially horrific, recent case in point. But a politically sophisticated electorate will not so easily be fooled; an electorate educated in wisdom will, in particular, demand of its government that it treats its electorate as adults, speaks the truth about economic and social realities, and implements wisdom-policies wherever feasible. Wisdom cannot be imposed on people from above; nor can it be imposed on government by the electorate. A wise society becomes possible only when those in power, and those who are governed, share a certain level of political sophistication and wisdom. But wisdom does not drop out of the air, as if by some miracle. It needs to be learned.

Knowledge based education and inquiry does not, in itself, help one acquire wisdom, not even political wisdom. Wisdom based education would however be designed to do just that. Knowledge-education cannot teach political wisdom: judged from the standards of knowledge-inquiry, any attempt to do that would degenerate into political indoctrination of one kind or another. The nearest one can get to political education is to study the political constitution of one's country, the political manifestos of political parties, or recent political history, since this involves the acquisition of factual knowledge, and not the making of political judgements. Granted wisdom-education, however, imaginative exploration of problems of living, imaginative and critical exploration of possible solutions become central to the whole of education. In particular, critical examination of political doctrines and manifestos, critical examination of government policy and action, become an important part of education. The tendency of governments to deceive, to massage public opinion with propaganda, to incite hatred of traditional enemies, or even to go to war, in order to gain popularity: consideration of these and other such standard ploys of politicians and governments would form a basic part of wisdom based political education. At the same time, however, such education would indicate the danger, the self-defeatism of disillusionment with and disengagement from politics. Compromise, lost opportunities, muddle, politicians pursuing personal power, status or wealth at the expense of the public good: these are inevitable features of democratic politics in the real world, and do not constitute good grounds for turning one's back on the political scene. In these ways, wisdom-education is designed to promote political enlightenment, something that knowledge-education cannot do.

There is an important, more general, more fundamental way in which wisdom-education would help us discover how to live civilized lives — something which knowledge-education cannot provide. At the centre of wisdom-education, from the age of five (let us suppose) onwards, there would be a discussion seminar, concerned to encourage children to engage in the activity of articulating and scrutinizing problems and their possible solutions.[10] This seminar would be conducted in such a way as to encourage open-ended, uninhibited discussion, there being no prohibition on what problems can be discussed, what solutions considered. War, sex, death, power, the nature of the universe, money, politics, fame, pop stars, parents, school, work, marriage, the meaning of life, evolution, God, failure, drugs, love, suffering, happiness: whatever it is that the children find fascinating or disturbing, and want to discuss, deserves to be discussed. Where there are no known or agreed answers, the teacher must acknowledge this. The teacher must readily acknowledge his or her own personal ignorance or uncertainties, in addition to confessing his or her convictions. The main task of the teacher will be to try to ensure that the children speak one at a time, that everyone gets a chance to speak, and that those who are not speaking listen. The teacher will also, of course, try to establish a spirit of generosity towards the ideas of others, while at the same time encouraging criticism and argument. The main object of the seminar is to enable children to discover for themselves the value of cooperative, imaginative, rational problem-solving by taking part in it themselves. Only good, experienced teachers could hope to make a success of the seminar, run along these lines.

The purpose of the seminar is not to promote mere *debate*. Argument is to be used as an aid to exploration and discovery: it is not to be used merely to trounce opponents or to win converts — as an excuse, that is, for intellectual duelling or bullying. The seminar must not be conducted in such a way that it amounts to overt or disguised *indoctrination* in some creed — however correct or noble that creed may be judged to be. In so far as a creed is implicit in the seminar, it might

---

10. This seminar is of course the straightforward result of applying rules 1 and 2 of rational problem-solving (see chapter three, section 3.4) to education.

be put like this: it is proper and desirable for people to try to resolve problems and conflicts in cooperative, imaginative and rational ways. This creed is itself open to discussion and critical assessment — along with all other political, religious, moral, economic, social and philosophical doctrines. The problem of how to distinguish cooperative discussion from indoctrination deserves itself to be discussed when it arises. Again, the seminar is not *group therapy*. Its primary aim is not to solve urgent, practical, personal problems of the participants (although it may occasionally and incidentally do this). Problems can be imagined and do not need to be lived. Ideas can be aired as possibilities, and do not need to be believed. Accounts of personal experience are to be welcomed when relevant to the discussion, but are not expected or demanded. The aim of the seminar is to explore possibilities, and not to reach decisions about actions. Unanimity does not need to be sought. One would expect the seminar, however, to feed into, and be fed by, other parts of education: science, literature, history.[11]

The hope would be that wisdom-education, conducted along these lines, would achieve three things not achieved by knowledge-education.[12] First, it would help pupils to discover that rational inquiry, the sciences and the humanities, or culture more generally, are there to be used by the individual to enhance his or her own capacity to realize what is of value in his or her own life. The sciences and the humanities are the outcome, the record, of individuals (more or less cooperatively) searching, striving to solve problems of knowledge and understanding, problems of realizing what is of value; it is there for us to use in our search for what is of value, and to contribute to, ourselves, as best we can. Second, wisdom-education would produce a politically enlightened electorate. And third, it would help produce people who are able to, and desire to, tackle life's problems and conflicts in cooperatively rational ways, to the extent that this is possible.

In the many different ways indicated, then, wisdom based academic inquiry would have a profound effect on the way people and institutions think about and tackle problems of civilization.

---

11. This feature of wisdom-education amounts to a straightforward application of rules 3 and 4 of rational problem solving (see chapter three, section 3.4).

12. Wisdom-inquiry would require there to be a version of this seminar at all levels of education, right up to the level of postgraduates and professors.

4.6 Apathy

At this point a further serious objection must be considered. One main obstacle to the creation of a more civilized world, it may be argued, is simply public apathy towards any such project. The goal is too distant, too vast, too abstract and impersonal to excite personal excitement and involvement. University based wisdom-inquiry might be able to bring to the world's attention the serious problems of civilization; it might be able to specify a number of severely scrutinized policy options which, if adopted, really would help the world become more civilized. But unless wisdom-inquiry engages in some form of indoctrination or brain-washing, unless it manages to inspire and inflame, the rational act of pointing to problems and possible solutions, possible policies, is all too likely to be greeted with a public yawn. People are too embroiled in the immediate problems of their lives, problems of careers, earning a living, bringing up and providing for a family, to take seriously great public measures designed to bring about a more civilized world. Especially will this be the case if these measures involve some degree of sacrifice of time, energy or money.

There is, I believe, in most of us a profound and passionate hunger for a better world, a world free of horrors suffered by so many in the 20th century. Most of us have a deep interest in the creation of a more civilized world. This is obvious as far as the poorest fifth of the world's population are concerned. But it is true also of those who live in wealthy, industrially advanced countries. Most of us have children, or hope to have children, and hope to have grandchildren, even great grandchildren. Those of us who are childless nevertheless are involved with the lives of our spouses, friends or family, or involved with pursuits, institutions, clubs or societies. We cannot help but hope that something of what we care for will survive our own personal death. If we have children, we must be concerned about what kind of world that they, or their children, will inherit. We must have an interest in the future flourishing of humanity, its capacity to avoid the destructive horrors of modern war, poverty, totalitarianism.

One reason for widespread apparent apathy comes from a deep suspicion of all political programmes or ideologies that promise to deliver civilization on earth. In the past, humanity has so often been bamboozled and betrayed by false promises. The Enlightenment led to the gruesome nightmares of the French revolution. Socialism and Marxism led to the Russian revolution, to Lenin, and then to Stalin, to the Chinese revolution, Mao and the cultural revolution. Right wing political ideology has produced Franco, Mussolini, Hitler, and any number of other dictators since. Promises of Utopia are pure fantasies which, if taken seriously lead, not to universal happiness but to war, repression, tyranny, arbitrary imprisonment, torture and death.

The New Enlightenment approach to world civilization differs from all of the above in countless ways. It is based on cooperative rationality and democracy, not on force (force being used only to restrain criminality). Wisdom-inquiry sets out to put forward and scrutinize possible policies for a better world, so that political programmes and actions may be based on well-tested ideas, and not on the inspiration and charisma of political leaders. Academics have no legislative power; they can only propose and criticize, and learn from the non-academic world. If wisdom based academic inquiry really does come up with practical, worthwhile proposals which, when implemented, really do help solve problems of civilization, people may come to believe that gradual social progress towards global civilization is a possibility (even if they do not lose a healthy scepticism).

Apathy can also come from impotence. What can the individual hope to accomplish, confronted by the vast, complex, impersonal modern world? When people lived in isolated hunting and gathering societies, each individual had a political voice, some chance of influencing the life of the tribe. Since those far off days, the tribe has become the human race, some five-and-a-half-billion people in number, all but a minute fraction complete strangers, living unfamiliar lives in far off places, speaking incomprehensible languages. No wonder the individual today feels powerless, helpless, apathetic.

An important function of wisdom-inquiry is to respond to this malaise of powerlessness and apathy. When a serious problem confronted the tribe, in the old days of hunting and gathering societies, it was at least possible for the tribe to sit down by the fireside and explore possible responses. Today this is not logistically possible: the tribe is too big. We need an institutional substitute for tribal discussion. It is an important part of the function of wisdom-inquiry to provide such an institutional substitute.[13] Wisdom-inquiry needs to provoke and sustain cooperatively rational tackling of problems of living in highly local, inter-personal ways; and it needs to inter-relate this with global thinking about global problems. The abstract, impersonal aspects of wisdom-inquiry are there partly as a reflection of the impersonal aspects of the universe, partly to accommodate thought about millions, or even billions of people. Wisdom-inquiry needs, however, to interrelate the impersonal and the personal, so that individual persons are encouraged in their personal thinking to come to grips with the impersonal — the vast, complex social world of humanity, the somewhat vaster world of the cosmos. By inter-connecting the local and the global, wisdom-inquiry would help individuals discover how to act locally in such a way as to help local conditions to flourish without detriment to the world as a whole — so that local consumption of food, drink or fuel, for example, does not cause pollution, poverty or extinction of species elsewhere on the planet. As a result of making very public what can be done locally to help globally, wisdom-inquiry would help people to acquire power to affect beneficially their own lives and the lives of people immediately around them, without at the same time contributing to global damage, and even perhaps making a minute, individual contribution towards the creation of a better world. With wisdom-inquiry of this type in place, it would be possible for individuals to come to terms with what they, as individuals, can achieve globally, especially when individual action can be coordinated with the action of others.

From what I have said so far, it may seem that I hold that, in order to build a better world it suffices to point out to the world's population (a) that it is in all our long-term interests to build a better world, and (b) what it is that individuals need to do to bring about a better world. The idea, here, is that once individuals appreciate (a) and (b), they will go

---

13. The primary substitute is representational government. But in addition to deliberations associated more or less directly with the execution of political power, we need sustained deliberations carried on openly in public, concerning matters not always on the immediate political agenda, deliberations removed from political power, and thus free of the distortions brought about by power. Once upon a time the priesthood might have fulfilled this role. Today, neither the civil service nor the press can fulfil the role. Knowledge-inquiry cannot do it either, but wisdom-inquiry is designed specifically to take it up. Wisdom-inquiry might include shadow governments seeking ideal policies.

ahead and do what is specified in (b) to bring about (a).

This is, of course, for all sorts of reasons, ludicrously naive. A "better world", as we have characterized it, is very much more in the interests of the poor than the wealthy. Even if the poor are eager to make a start on building a better world, the wealthy and powerful may not be so keen, and may well arrange matters so as to make it impossible for the poor to act, or even inform themselves about what needs to be done. But quite apart from this naivety (which all traditional socialists will pounce upon), there is another, even greater naivety in the above scenario. It expects individuals to act in new ways, for the sake of the general good, in the hope that others too will act in these new ways, merely because it has been determined that if this is done, everyone will benefit.

As a matter of fact people do act in such quasi-altruistic ways.[14] People sort out their rubbish into different categories, and take bottles to bottle banks, in the hope that others too will do this so that the environment may be protected. Nevertheless, if we are to make progress towards global civilization we cannot, I believe, rely on people taking up such new quasi-altruistic actions as our main strategy.[15] Instead we need to create conditions, by legislation, taxation and in other ways, which are such that it is in people's own interests to adopt actions conducive to promoting civilization. Democratically and rationally, we need to manipulate ourselves so that our circumstances become such that we desire and seek what we ought to desire and seek from the standpoint of creating civilization.

4.7 Inefficacy of Academia

Another objection that may be made to what has been said in this

---

14. I say 'quasi-altruistic' because what is involved is not acting for the benefit of others *per se*, but rather, acting for the benefit of oneself and others in the expectation that others will do likewise.

15. I say "new quasi-altruistic actions" because all of social life, in a sense, relies on already established quasi-altruistic actions. What is more questionable is that *new* quasi-altruistic actions will become established in a society once the need for such actions has become generally recognized. Contemporary life exhibits many counter-instances. Everyone agrees in Britain, at the time of writing, that there are too many cars on the road; everyone expects everyone else to cut down on their use of the car. Few unilaterally make less use of their car in the expectation that others will follow.

chapter so far (touched on above) is that it attributes to academia a degree of social influence that is far in excess of what it is capable of. Far from being in the thick of political life academics, typically and notoriously, live secluded lives, devoting themselves to the study of esoteric matters of no relevance to the lives of ordinary people. The very word "academic" has acquired the secondary meaning "useless", "irrelevant", "pointless".

But even the harshest critics of academia (on the grounds of its uselessness) must concede that parts of academic inquiry have had an immense impact on the modern world. Science and technology have transformed the world we live in, and scientific and technological research are largely, though not exclusively, a part of academic inquiry. (This is especially true of the more fundamental research.) As I have already remarked, universities exercise an impact on the social world via education, and via their training of professionals: doctors, engineers, lawyers, teachers, and so on. And in a multitude of ways, academics today, working within the framework of knowledge-inquiry, influence the social world around them by means of publications, television, official reports, contacts with government and industry.

The crucial point, in the present connection, however, is that the transition from knowledge-inquiry to wisdom-inquiry changes dramatically the role that universities have in society, and hence the impact that universities have on social life.

The biggest change, in moving from knowledge to wisdom inquiry, is a change in the basic intellectual *aim* of inquiry. Whereas for knowledge-inquiry the intellectual aim is acquisition of knowledge, which needs to be sharply dissociated from the social aims of inquiry, for wisdom-inquiry the intellectual and social aims are the same: to promote wisdom in life. This difference dramatically affects the whole way in which academia seeks to relate itself to the rest of the social world.

Given knowledge-inquiry, the primary task of the academic (scientist or scholar), is not to engage with the social world outside academia; it is rather to contribute to academic knowledge and scholarship. Eventually, no doubt, new knowledge will be taken up and used, where relevant, by those outside academia, by industry, the military, government, the general public. Any such use of new knowledge is not, however, the primary concern of the academic, which is to add to the store of academic knowledge, whatever its subsequent human use or value may turn out to be. In a secondary way, of course, academics may write popular works intended for the general public; they may write for newspapers or take part in TV or radio programmes; and they may do consultant and other work for industry or government. But, in line with philosophy-of-knowledge standards, such work tends to be regarded by academia as somewhat suspect, adding nothing to the scientific or scholarly reputation of the academic involved. (But attitudes here may be somewhat hypocritical, a cover for envy.) Again, academics may become involved with environmental or civil rights pressure groups: such work is quite distinct from, and adds nothing to the proper, purely academic work of contributing to knowledge. Granted knowledge-inquiry, in other words, the primary task is to make contributions to knowledge internally, within academia itself: it is not to engage with the outside social world, and cannot be to take part in some morally committed political campaign. In acting thus the academic ceases to act as scientist or scholar and becomes an ordinary citizen like any other.[16]

Given this, and granted that knowledge-inquiry by and large prevails today, it is not surprising that most people outside universities should regard what goes on inside universities as largely irrelevant to the big political, economic, social issues of the day. It is not surprising that the main thesis of this book — in order to build a better world we need to free science, and academic inquiry more generally, of their neuroses — is likely to be greeted with some scepticism. How can changing something as "academic" as academic inquiry change anything that really matters?

But granted wisdom-inquiry, the way in which academia is related to the rest of the social world would change dramatically. The basic intellectual aim of inquiry is the same as the social aim: to promote wisdom in life. Contributions to thought are a means to that end. Thus, from the standpoint of wisdom inquiry, the primary, the fundamental task of the academic is to speak to, to engage in discussion with, people

---

16. In recent times in the UK there has been a tendency for academics increasingly to work for industry and government; this makes it all the more important that wisdom-inquiry standards operate within universities, guiding involvement of universities with the rest of society for the good of humanity, and not just for the good of universities, the wealthy and powerful.

outside universities. Writing newspaper articles, giving public lectures, taking part in TV programmes, writing books and articles for public consumption, far from being dubious non-professional activities, at best secondary to the main academic business of contributing to knowledge, are on the contrary the central academic concern. Educating one's fellow academics by publishing scientific or scholarly articles in academic journals, even though important, is nevertheless a means to the end of educating the public. What ultimately matters is public wisdom, and not just wisdom of the professors.

The *content* of what academics have to communicate changes dramatically as well, as we move from knowledge to wisdom inquiry. Instead of the products of inquiry being exclusively contributions to knowledge, they become, in addition, expressions of problems of living, and proposals for their solution: economic and political policies, criticisms of current actions and policies, political philosophies and philosophies of life. Wisdom-inquiry is directly concerned with ideas implicated in the way people live in a way in which knowledge-inquiry cannot be. It would, in short, be impossible to dismiss wisdom-inquiry as merely "academic".

At this point defenders of the academic *status quo* may object that putting the philosophy of wisdom into practice would have the effect of destroying all genuine science and scholarship and transforming academia into nothing more than another campaigning organization.

But this objection fails to grasp the nature of what is being proposed. The central task of wisdom-inquiry is to help humanity acquire wisdom (by cooperatively rational means). Wisdom includes knowledge, technological know-how and understanding. Science and scholarship are thus vital to wisdom-inquiry. A part of the case for the philosophy of wisdom, indeed, is that wisdom-inquiry does far better justice to the intellectual, the cultural value of science and scholarship pursued for their own sakes than does knowledge-inquiry.[17]

---

17. This point was made in section 3.4 of chapter three: see, too, section 2.3 of chapter two, and pp. 125-127 and 138-140 of the present chapter. See also (Maxwell, 1984a, pp. 181-189; 1976a, pp. 98-102 and 138-157).

4.8 Future Prospects

Much of my working life has been taken up with developing and attempting to communicate the message of this book, namely: there is an urgent need to free universities of their neuroses so that they take up their proper task of promoting wisdom by rational means. My first (published) attempt was made in 1976 with the appearance of *What's Wrong With Science?*. Naively, I thought the book would work its magic, and scientific neurosis would begin to crumble away. Needless to say, nothing of the kind happened. In those far off days the world was in the grip of the Cold War and the nuclear arms race. There was some awareness of other impending global problems, in part as a result of the publication of such books as Rachel Carson's *Silent Spring* (1965), Barry Commoner's *The Closing Circle* (1971), the Club of Rome's *The Limits to Growth* (1972), Schumacher's *Small is Beautiful* (1973), and Ronald Higgins's *The Seventh Enemy* (1978). Although most of the authors of these books were academics, nevertheless academia itself seemed in those days so constituted that one would never expect it to support and sustain work of this kind. In the 1960s and 70s the neuroses of science, and of academic inquiry more generally, were much more rigidly maintained than they are, perhaps, nowadays in the first decade of the 21st century. *Then* one could not, as a normal part of academic work, publish a book warning of dire problems facing humanity, especially if this involved the use of modern science and technology. Rachel Carson's *Silent Spring* provoked a storm of protest from fellow academic chemists. Even among the public, there was nothing like the suspicion of modern science, technology and medicine that one finds today.

Since those times, much has changed, some changes for the better, some for the worse. The Cold War is at an end, but so many of the opportunities that the collapse of the Soviet Union provided us with have been squandered. The USA and Europe failed to give substantial help to Russia to become a prosperous liberal democracy, and instead gave bad economic advice. Third world debt was not cancelled. There was no new Marshall plan to help the third world become more prosperous and democratic, and less corrupt. Instead we have had the horrific wars associated with the disintegration of Yugoslavia, massacre in Rwanda and elsewhere in Africa, and the continuing horror of the conflict between Israel and Palestine. And we now suffer from international terrorism, the threat of terrorism, and perhaps even worse, the consequences of the misconceived "war against terrorism". Bin Laden's purpose, presumably, in supporting the attack on the twin towers in New York in September 2001, was to provoke a war between Islam and "the West". If so, Bush and Blair fell straight into his trap, in going to war with Iraq in March 2003. It is almost as if Bush is doing just what bin Laden wants him to do.

Since the 1970s, there has been a massive increase in public awareness of environmental problems. There is far greater distrust of modern science, and of experts of all kinds. Along with growing distrust of conventional medicine, there has been an increase in alternative medicine of many varieties. These changes in public attitudes are often associated with Romantic, anti-rationalistic ways of thinking.

And there have been many changes in universities, some of which may be interpreted as the gradual, piecemeal implementation of aspects of wisdom-inquiry. Research into environmental problems has become a standard part of academic work. Senior scientists concern themselves with the public understanding (and misunderstanding) of science. Issues concerning public participation in science, and the democratic control of science are increasingly discussed. Centres of interdisciplinary studies and policy studies have been created. The significance of Darwinian theory for social inquiry, for philosophy and philosophy of science has come to be much more widely appreciated (although the Darwinian point, stressed at the end of section 3.5 of the last chapter, that human learning should be seen as a rational development of animal learning, having as its basic task to solve problems of living, has not yet been appreciated). Increasingly, the social and moral aspects of science and technology are scrutinized. Programmes that seek to "teach for wisdom" have been set up in the USA. Academics have become much more aware of the importance of communicating with the public. A spate of books has appeared popularising aspects of science.

From the sidelines, I have watched these changes come about with a mixture of hope and exasperation. Hope, because these changes can indeed be seen as steps towards wisdom-inquiry. Exasperation, because it all seems to be happening in such a piecemeal, muddled and limited way — and in complete ignorance, of course, of my own attempted contribution to the revolution so slowly taking place. Some of those involved in some of the changes just indicated oppose what they see as the threat of the spreading influence of scientific rationality — not appreciating that what they oppose is scientific neurosis masquerading as rationality. On the other hand, there are many scientists who view some of the changes I have indicated as posing a severe threat to science itself, and to our whole way of life, not appreciating that what they seek to preserve is scientific neurosis, not scientific rationality. Philosophers of science are at odds with historians and sociologists of science, and those who work in the humanities, more generally, often fail to see that their work has anything in common with the work of scientists or mathematicians. Changes being attempted in one department of academia seem to have nothing in common with changes being attempted elsewhere. It all lacks focus and clarity of purpose. And because academics are no longer convinced academia has an overall purpose to which diverse specialities contribute in a coherent way, their loyalty goes, not to academia as a whole, but to their own particular research, their own particular speciality or department.

My hope is that this book will help bring a certain impetus and focus to changes already going on, and will help academics in different specialities find a common purpose so that, working together across disciplines, they bring about the intellectual revolution we need, and academia takes up its proper task of promoting wisdom in the world.

Who can doubt, in view of our record of recent mismanagement of our affairs, and the dangers we face, that we urgently need to acquire a little more wisdom?

Appendix

In this appendix I discuss some slightly more technical issues arising, in the main, from the arguments of chapters one and two concerning the rationalistic neurosis of theoretical physics.

The issues discussed are, in order, the following:

(1) Arguments designed to establish that science makes metaphysical assumptions about the universe.
(2) Problems concerning simplicity and unity of physical theory.
(3) Arguments for and against aim-oriented empiricism.
(4) Rational discovery of new, fundamental physical theories.
(5) A possible alternative to physicalism.
(6) Whether aim-oriented empiricism solves the problem of induction.

1 Physics and Metaphysics

In chapter one I argued that persistent acceptance of theories that bring greater overall theoretical unity to physics, when endlessly many empirically more successful, but disunified, rival theories exist, implies that science makes a persistent metaphysical assumption about the universe: it is such that there is some kind of underlying unity in nature, no disunified theory being true. This is, in my view, a decisive argument, providing decisive grounds for rejecting standard empiricism. Most contemporary philosophers of science, however, ignore the argument, ignore the outcome of the argument, aim-oriented empiricism, and continue to take one or other version of standard empiricism more or less for granted, even though the argument has been in print for thirty years at the time of writing: see (Maxwell, 1974). The problem of the "underdetermination" of theory by evidence is generally understood to be a fundamental problem confronting the philosophy of science, but the implications of underdetermination (the need to make explicit implicit metaphysics, and thus the need to reject standard and accept aim-oriented empiricism) remain, for most, hidden from view.[1] A striking

---

1. Philosophers of science manage to hold onto one or other version of standard empiricism (despite its decisive refutation) by means of what C. P Snow, in another context, called "the technique of the intricate defensive" (Snow, 1964, p. 67). As I have

example of this is provided by James Brown in his recent book *Who Rules in Science?*. During the course of his book, Brown makes some excellent points in criticism of social constructivist and sociologists of science, but then throws it all away with his lamentably inadequate discussion of "underdetermination", as he calls the problem (Brown, 2001, pp. 162-166). Brown asks why empirically successful rivals to accepted theories deserve to be rejected. His answer is that in practice there are only very few rival theories; choosing between them can therefore be done by an appeal to evidence — and that, he declares, solves the problem. If this is the best those who defend the rationality of science can do, then no wonder so many argue that interests and social factors determine what theories are accepted, scientific knowledge thus amounting to no more than contemporary myth. What Brown's answer ignores, of course, is that, as we saw in chapter one, endlessly many empirically more successful rivals can always be concocted, and infinitely many exist. These don't disappear just because they are ignored in scientific practice. Scientists' persistent (and sensible) neglect of almost all of these empirically more successful rivals means that science makes a persistent (metaphysical) assumption about the universe. Precisely because of the highly influential and problematic character of this assumption, it needs to be made explicit within science, so that it can be critically scrutinized, so that alternatives can be considered, in the hope that this will lead to the assumption being improved. It is this line of argument that leads to the rejection of standard empiricism and the acceptance of aim-oriented empiricism instead. Aim-oriented empiricism makes explicit, and subjects to critical scrutiny, all the factors "determining" choice of theory — evidential and metaphysical — thus exhibiting science as a rational endeavour, and leaving no room for social constructivists and sociologists of science to argue that social factors external to science must play a role in deciding what theories are accepted and rejected. (Such social factors do, of course, influence priorities of scientific research.)

Brown does at least formulate the "underdetermination" problem properly, even if his attempted solution is hopeless. Others formulate it

---

put it elsewhere (Maxwell, 1984a, p. 35), "Discussion of the problems confronting standard empiricism . . . becomes so elaborate, technical and abstruse, that the simple and decisive objections to the position are lost sight of by everybody, and the position is preserved by default".

in such a way that the problem of rival theories is, as it were, all but defined out of existence. The problem of underdetermination is taken to be the problem of how one chooses between rival theories that agree as far as *all* empirical consequences are concerned: see (Kukla, 2001), and further references given therein. This version of the problem excludes from the outset the far more serious problem of how to choose between theories that agree as far as all *observed* empirical consequences are concerned, but differ for as yet unobserved consequences.

In the main, most philosophers of science remain unaffected in their work by the above refutation of standard empiricism leading to aim-oriented empiricism by the simple strategy of ignoring it. Thus a recent comprehensive survey of views concerning scientific method (Nola and Sankey, 2000) makes no mention at all of the argument, or of its outcome, aim-oriented empiricism.

One contemporary philosopher of science who has noticed the argument, if only to contest it, is Dr. F. A. Muller of Utrecht University. After a long exchange of emails, even Muller was brought round to conceding the argument's validity when, under pressure, I formulated it as follows.

In accepting a non-*ad hoc*, unified theory, T, the physics community thereby rejects infinitely many, easily formulatable, rival, *ad hoc*, disunified theories (such as those considered in section 2 of chapter one) which clash with T, and which are more empirically successful than T. Accepting T is choosing T from the available rivals: in accepting T, the rivals are thereby rejected. Accepting T, in other words, involves accepting:

(A) "Not T_1, and not T_2, and .... and not T_∞", where these are all empirically more successful *ad hoc* rivals to T.

Is (A) testable? The empirical verification of any one of T_1, T_2,... would falsify (A). But theories can't be verified. The falsification of all of T_1, T_2,... would verify (A). But as there are infinitely many of them, this can't be done either. So (A), being neither verifiable nor falsifiable, is metaphysical. In accepting T the physics community thereby accepts the metaphysical thesis (A).

It may be objected that this is not quite what had to be proved, namely that acceptance of non-empirical *methods*, M, implies acceptance of a metaphysical thesis. M may be accepted, for example, in the absence of any available, acceptable T.

Suppose that this is the case, and M specify what it is for a theory to be unacceptably *ad hoc* (whatever the empirical success of the theory might be). Even though no theory, T, has been formulated which is both sufficiently empirically successful, and sufficiently in accord with M, to be acceptable, nevertheless there are still infinitely many easily formulatable, empirically successful *ad hoc* theories which are rejected — or rather, are not even considered — because they fail to accord with M. Acceptance of M involves the rejection of all these theories, T_1, T_2, ... T_∞; that is, it involves the acceptance of:

(B) Not T_1, and not T_2,... and not T_∞.

But, as before, (B) is neither falsifiable nor verifiable; hence it is metaphysical. Thus, acceptance of M, even in the absence of an empirically successful theory which accords with M, implies acceptance of a metaphysical thesis, namely the thesis (B).

A further decisive argument in support of the contention that physics makes the metaphysical presupposition that there is an underlying unity in nature is the following.

Deny that physics makes any metaphysical assumption — accept, in other words, some version of standard empiricism (SE) — and it becomes impossible to say what it *means* to declare of a theory that it exhibits (or fails to exhibit) unity. There have been numerous attempts to say what simplicity or unity mean within the confines of SE: see Jeffreys and Wrinch (1921), Popper (1959, ch. 7), Goodman (1972), Friedman (1974), Sober (1975), Kitcher (1981, 1989), Watkins (1984, pp. 203-213), McAllister (1996). All fail. For criticisms see Salmon (1989), Maxwell (1998, ch. 2) and (2004e). Even Einstein, one might note, recognized the problem, but declared that he did not know how to solve it, although he thought it ought to be solvable (Einstein 1949, p. 23).

There is a very simple reason why all such attempts must fail. Any attempt to characterize the structural simplicity or unity of a theory within the confines of SE has to do this in terms of the way the theory is *formulated*, or *axiomatized*. But a theory that seems beautifully simple and unified given one formulation or axiomatization, can always be reformulated, re-axiomatized, in endlessly many different ways, to come out as horribly complex or disunified as one wishes. And *vice versa*: any theory, however complex or disunified its formulation may be, can always be reformulated, re-axiomatized, so that it comes out as beautifully simple and unified. Richard Feynman (Feynman *et al.* 1965, pp. 25-10 – 25-11) has given a delightful example of how the latter can be achieved.

Consider an appallingly complex universe governed by 10^{10} quite different, distinct laws. Even in such a universe, the true "theory of everything" can be expressed in the dazzlingly simple, unified form: $A = 0$. Suppose the 10^{10} distinct laws of the universe are: (1) $F = ma$; (2) $F = Gm_1m_2/d^2$; etc. Let $A_1 = (F - ma)^2$, $A_2 = (F - Gm_1m_2/d^2)^2$, etc., for all 10^{10} distinct laws. Let: $10^{10} A = \sum A_r$. The true "theory of everything" of this universe can now be $r = 1$, formulated as: $A = 0$. (This is true if and only if each $A_r = 0$.)

How does AOE overcome this difficulty? By demanding that, in assessing the simplicity or unity of a theory, what must be attended to is the *content* of the theory, what it asserts about the world, and not the theory itself, as formulated or axiomatized in this or that way. Thus the difficulty that is fatal to all attempts at solving the problem within the constraints of SE, is no problem at all, granted AOE. Given any theory, formulated in a number of different ways, T_1, T_2, ...T_n, some extremely simple and unified, others horribly complex and disunified, but all specifically interpreted and designed to have the same content, make precisely the same assertion about the world, then all these formulations have precisely the same degree of simplicity or unity just because the content is the same in all cases.[2]

---

2. Whenever I have lectured on this solution to the problem of simplicity or unity, the objection has been made that it is not at all clear whether, given two versions of a theory, the two versions really do have the same content or not. But this objection entirely misses the point. The difficulty, which SE cannot solve, and AOE can, *presupposes*, in its very formulation, that we are confronted by a number of different formulations of a theory, some unified, some not, but all making the same assertion about the world. Such diverse formulations of one and the same theory can always be concocted artificially, by introducing special terminology, as Feynman beautifully illustrates. The fact that, on occasions, we do not know whether two formulations of a theory have precisely the same content or not is a complete red herring. The crucial point is that, given any theory whatsoever, it is a trivial matter to concoct diverse formulations, some complex, some dazzlingly simple, which are deliberately concocted to have the same content. It is this which creates the insuperable difficulty for SE.

But why, it may be asked, cannot SE avail itself of exactly the same solution? The answer is that if science adopts a methodological rule which stipulates that all theories whose *content* lack such and such a degree of unity must be rejected, whatever their empirical success may be, then science thereby in effect accepts, as a part of scientific knowledge, the metaphysical thesis that the universe is such that all theories which fail to have the required degree of unity are false. And this violates SE.

But could not a modified version of SE be adopted which acknowledges that metaphysical theses are accepted by science but which, nevertheless, insists that these theses are accepted and rejected solely on empirical grounds? Metaphysical theses cannot be assessed directly on empirical grounds, since they are, by definition, neither verifiable nor falsifiable empirically. They can however be assessed in terms of the empirical success or failure of the research programmes to which they give rise. In order for a research programme, R, based on a metaphysical assumption, M, to meet with empirical success, two requirements must be met. First, R must generate a succession of theories which progressively more and more successfully capture M as a precise, testable theory. Second, the succession of theories must meet with ever greater empirical success. Given two rival research programmes, R_1 and R_2, based on rival metaphysical theses, M_1 and M_2, R_1 may be deemed to be preferable to R_2 on the grounds that it meets with greater empirical success. In choosing between rival research programmes in this way, in terms of relative empirical success, we also choose between rival metaphysical assumptions, M_1 and M_2, on empirical grounds.[3]

The modified version of SE under consideration, then, holds that the metaphysical assumption, M, of science, at any given stage in its

---

3. This is similar to, but not precisely the same as, the picture of scientific progress developed by Lakatos (1970). The main difference is that, for Lakatos, the "hard core" of a research programme was a testable theory rendered metaphysical by a methodological decision; the main research activity associated with a research programme involved developing successful applications of the theory, guided by the "positive heuristic" stemming from the "hard core". (In all this, Lakatos followed Kuhn's conception of "normal science", giving Lakatosian terms to Kuhnian ideas: see (Kuhn, 1970).) In the text, I have assumed that the metaphysics of a research programme is authentic, untestable metaphysics, the main research task being to develop a succession of theories which progressively capture the metaphysics more and more successfully.

development, is selected solely on empirical grounds, in the way just indicated, M changing as science progresses, and old research programmes die and new ones are born. Methodological rules governing simplicity or unity stipulate that the *content* of physical theory must, as far as possible, be compatible with the current M, these rules also changing as science progresses.[4]

There are at least four objections to this Lakatosian version of SE (LSE).

First, almost all physical theories put forward so far, from Newtonian theory to quantum theory, can be regarded as failing to satisfy some requirement of unity, so that they fail to be perfectly explanatory, even from their conception. Newtonian theory leaves unresolved mysteries having to do with inertial mass, the equality of inertial and gravitational mass, the role of space and the absolute nature of acceleration. Classical electrodynamics fails to solve the problem of how point-charges can both create and be acted upon by the field. Orthodox quantum theory is, as I indicated in chapter two, severely disunified. Even general relativity is unsatisfactory, in that it cannot properly accommodate matter, and predicts that black holes must exist, the singularity of a black hole constituting a breakdown of the theory. But LSE cannot account for these unity defects of existing theories since, according to LSE, it is these theories, when empirically successful, which set the standard of what is to count as "fully unified". LSE cannot make sense of the idea that accepted, empirically successful theories are defective from the standpoint of unity, even when first accepted, since it is just such a theory which represents the ideal of unity, according to LSE.[5]

Second, LSE can give no account of the discovery, or the creation, of

---

4. For a conception of simplicity or unity somewhat like the SE view just indicated, see (McAllister, 1996). The main differences are that McAllister does not depict scientific progress in the somewhat Lakatosian way that I have done, and does not state clearly that it is the *content* of theories that is crucial when it comes to assessing their simplicity or unity — or "beauty". Instead, McAllister, rather unsatisfactorily, speaks of theories construed as abstract intellectual entities. For a critical comparison of McAllister's version of SE and AOE see (Maxwell, 2004e).

5. Some of the criticisms of theories indicated have to do with consistency. LSE could be formulated to demand, *a priori* as it were, that a theory, in order to be acceptable, must be consistent. But this does not resolve the difficulty. All the theories considered can be formulated so as to be consistent: whenever inconsistency threatens, the theories are simply arbitrarily curtailed, so that potential inconsistencies are excluded. The result is a consistent but severely *ad hoc* theory.

new fundamental theories.   Such theories almost always contradict earlier theories that they supersede.   They almost always bring about a metaphysical revolution, as well as a theoretical one.   Thus Newtonian theory, in postulating the gravitational force which acts instantly at a distance, clashed   with   earlier   metaphysical   ideas.   Classical electrodynamics, in postulating the electromagnetic field, clashed with Newtonian ideas.   And the same goes for special and general relativity, quantum theory, quantum field theories, and string or M theory. LSE, unlike AOE, contains nothing which can provide a basis for the development of these new metaphysical and theoretical ideas which clash with earlier ideas.

Third, LSE is obstructively conservative.   As we have just seen, new fundamental theories, and associated metaphysical ideas, break with their predecessors; in order to develop and assess new theories properly, in terms of non-empirical criteria having to do with unity and explanatory perfection, new criteria are needed, associated with new metaphysical ideas.   It is just this which LSE cannot provide. According to LSE, new theories have to be assessed in terms of metaphysical ideas, and related requirements of unity, associated with past theories, precisely the *wrong* metaphysical ideas, the *wrong* requirements of unity. This can only serve to obstruct the progress of physics.

Fourth, LSE fails to solve the problem of how to exclude empirically successful but radically disunified theories, of the kind considered in chapter one.   For, corresponding to such highly empirically successful but grossly disunified theories (never considered in scientific practice for a moment), there will be grossly disunified metaphysical assumptions which could form the basis for highly empirically successful research programmes.   LSE is quite unable to provide grounds for rejecting such "disunified" theories, metaphysics, and associated research programmes. In fact, even worse, LSE is obliged to declare that such "disunified" metaphysical assumptions represent perfect standards of unity, since they are   associated   with   empirically   successful   theories   and   research programmes. This is a lethal objection to LSE.

The above difficulties pose no problem for AOE.   First, AOE assumes, at   level   4,   that   the   universe   is   perfectly   physically comprehensible. It is this which provides a basis for criticizing theories that fail to meet ideal requirements for being perfectly unified or explanatory.   Second, as I indicated in chapter two, AOE provides a rational, if fallible and non-mechanical, method for the discovery of fundamental new theories. Third, AOE provides requirements of unity that are extremely demanding and the very opposite of conservative. They will only be satisfied by a physical theory of everything that is perfectly unified and explanatory, to the extent, even, of unifying particles-and-forces on the one hand, and space-time on the other. Fourth, AOE solves the problem of excluding empirically successful but radically disunified theories without difficulty. Such theories clash with the level 4 thesis that the universe is physically comprehensible.

So far I have put forward three arguments in support of the claim that physics makes metaphysical assumptions, namely:

(1) Persistent acceptance of unifying theories, and persistent rejection of disunified but empirically more successful rivals commits physics to making the metaphysical assumption that that which determines how events unfold exhibits unity to some extent.
(2) It commits physics to making the metaphysical assumption: "Not T_1, and not T_2, and ... and not T_∞", where T_1, T_2, etc. are all radically disunified theories.
(3) Only a view which construes science as making a metaphysical assumption about some kind of underlying unity in nature can make sense of criteria of unity governing acceptance of fundamental theories in physics. Even LSE, a Lakatosian version of SE, which holds that metaphysical assumptions are to be selected solely on the basis of empirical fruitfulness, fails to explicate unity adequately despite acknowledging metaphysical theses associated with science, and is in any case untenable.

These three arguments do not, however, establish AOE. Even if it is granted that metaphysical assumptions concerning some degree of unity in nature are presupposed by scientific method (and are not selected solely on the basis of empirical fruitfulness as LSE would have it), nevertheless these assumptions might be weaker than those specified by AOE. In the next but one section I will consider what further arguments there are in support of AOE. But first I must discuss the capacity of AOE to solve problems concerning the simplicity and unity of physical theories.

2 Problems Concerning Unity and Simplicity of Physical Theory

AOE has, it seems, a straightforward account of what theoretical unity in physics means. Unity applies, in the end, to the whole of theoretical physics. The totality of fundamental physical theory, T, applicable in principle to all phenomena, is unified if its *content* is the same throughout all phenomena — that is, if the dynamical laws asserted by T are *the same* throughout all phenomena.[6] What matters, as I have already stressed, is not the way T is formulated, but its content, what T asserts about the world. Given any number of different formulations of T (which can always be devised given one formulation), some beautifully simple and unified, others horribly complex and disunified, as long as they all have the same content they all have the same degree of simplicity or unity.

Furthermore, degrees of unity are to be characterized like this. If N different laws apply in N different ranges of phenomena to which T applies, then T is disunified to degree N (with unity arising if N = 1).

This account of unity (and the elaborations to be discussed in a moment) can be extended to apply to individual physical theories that are not theories of everything in two ways. First, given two rival theories, T_1 and T_2, and given the rest of fundamental physical theory, T* (which would include mere laws if a theory for some range of phenomena is lacking), then we can declare that T_1 is more unified than T_2 if T* + T_1 is more unified than T* + T_2. Second, we can just apply the above conception of unity to the individual theory — treating it, in effect, as if it is a theory of everything. This second option is adopted in what follows.

To give an elementary example, Newton's theory of gravitation, $F = GM_1M_2/d^2$ is unified in that what the theory asserts is *the same* throughout all possible phenomena to which it applies (all bodies of all possible masses, constitution, shape, relative velocity, distance apart, at all times and places). An *ad hoc* version of this theory, which asserts that $F = GM_1M_2/d^2$ for times $t \leq t_0$, where t_0 is some definite time, and

---

6. If T is formulated in terms of a set of differential equations then it is what these equations assert that must be the same throughout the diverse phenomena that T postulates as physically possible. Laws specifying precisely how diverse physical states evolve in space and time may be quite diverse in character: what matters is that they are all solutions of the same set of differential equations.

$F = GM_1M_2/d^3$ for times $t > t_0$, is disunified because what the theory asserts is *not* the same throughout the range of possible phenomena to which the theory applies.

Note that special terminology could be introduced to make Newtonian theory look *disunified*, and the *ad hoc* version of Newtonian theory look *unified*. All we need do is interpret "d^N" to mean "d^N if $t \leq t_0$ and d^{N+1} if $t > t_0$". In terms of this (admittedly somewhat bizarre) terminology, the disunified theory has the form "$F = GM_1M_2/d^2$" and Newtonian theory has the "disunified" form "$F = GM_1M_2/d^2$ for times $t \leq t_0$ and $F = GM_1M_2/d$ for times $t > t_0$". But this mere *terminological* reversal of unity and disunity does not affect the *content* of the two theories: the content of Newtonian theory remains unified, and the content of the *ad hoc* version (which looks unified) remains disunified.

This account of unity needs to be refined further. In assessing the extent to which a theory is disunified we may need to consider *how* different, or *in what way* different, one from another, the different contents of a theory are. A theory that postulates different laws at different times and places is disunified in a much more serious way than a theory which postulates the same laws at all times and places, but also postulates that distinct kinds of physical particle exist, with different dynamical properties, such as charge or mass. This second theory still postulates *different* laws for different ranges of phenomena: laws of one kind for possible physical systems consisting of one kind of particle, and slightly different laws for possible physical systems consisting of another kind of particle. But this second kind of difference in content is much less serious than the first kind (which involves different laws at different times and places). In other words, the more *different*, one form another, the different contents of a theory are, so the more seriously *disunified* the theory is.

What this means is that there are different *kinds* of disunity, different *dimensions* of disunity, as one might say, some more serious than others, but all facets of the same basic idea. Eight different facets of disunity can be distinguished as follows.

Any dynamical physical theory can be regarded as specifying a space, S, of possible physical states to which the theory applies, a distinct physical state corresponding to each distinct point in S. (S might be a set of such spaces.) For unity, we require that the theory, T, asserts that *the same* dynamical laws apply throughout S, governing the evolution of the physical state immediately before and after the instant in question. If T postulates N distinct dynamical laws in N distinct regions of S, then T has disunity of degree N. For unity in each case we require that $N = 1$.

(1) T divides spacetime up into N distinct regions, $R_1...R_N$, and asserts that the laws governing the evolution of phenomena are the same for all spacetime regions within each R-region, but are different within different R-regions. Example: the *ad hoc* version of Newtonian theory (NT) indicated above: for that theory, $N = 2$, in a type (1) way.[7]

(2) T postulates that, for distinct ranges of physical variables (other than position and time), such as mass or relative velocity, in distinct regions, $R_1,...R_N$ of the space of all possible phenomena, distinct dynamical laws obtain. Example: T asserts that everything occurs as NT asserts, except for the case of any two solid gold spheres, each having a mass of between one and two thousand tons, moving in otherwise empty space up to a mile apart, in which case the spheres attract each other by means of an inverse cube law of gravitation. Here, $N = 2$ in a type (2) way.

(3) In addition to postulating non-unique physical entities (such as particles), or entities unique but not spatially restricted (such as fields), T postulates, in an arbitrary fashion, $N - 1$ distinct, unique, spatially localized objects, each with its own distinct, unique dynamic properties. Example: T asserts that everything occurs as NT asserts, except there is one object in the universe, of mass 8 tons, such that, for any matter up to 8 miles from the centre of mass of this object, gravitation is a repulsive rather than attractive force. The object only interacts by means of gravitation. Here, $N = 2$, in a

---

7. As I have formulated it here, (1) is open to two somewhat different interpretations. First, for $N = 1$ we require only that *the same* law operates throughout space in the sense that this would be true even if the law in question asserted that all objects experience a force directed at a unique point in space, and inversely proportional to their distance from that point. Second, for $N = 1$, we require that *the same* law operates throughout space in the sense that a mere change of position in space of an isolated physical system has no effect on the way the system evolves. An analogous distinction arises in connection with time. In what follows I adopt the second interpretation of (1), which means that a theory which is unified with respect to (1) exhibits symmetry with respect to spatial location, and time of occurrence. As far as the *ad hoc* version of NT is concerned, $N = 2$ for both versions of (1).

type (3) way.

(4) T postulates physical entities interacting by means of N distinct forces, different forces affecting different entities, and being specified by different force laws. (In this case one would require one force to be universal so that the universe does not fall into distinct parts that do not interact with one another.) Example: T postulates particles that interact by means of Newtonian gravitation; some of these also interact by means of an electrostatic force $F = Kq_1q_2/d^2$, this force being attractive if q_1 and q_2 are oppositely charged, otherwise being repulsive, the force being much stronger than gravitation. Here, N = 2 in a type (4) way.

(5) T postulates N different kinds of physical entity,[8] differing with respect to some dynamic property, such as value of mass or charge, but otherwise interacting by means of the same force. Example: T postulates particles that interact by means of Newtonian gravitation, there being three kinds of particles, of mass m, 2m and 3m. Here, N = 3 in a type (5) way.

(6) Consider a theory, T, that postulates N distinct kinds of entity (e.g. particles or fields), but these N entities can be regarded as arising because T exhibits some symmetry (in the way that the electric and magnetic fields of classical electromagnetism can be regarded as arising because of the symmetry of Lorentz invariance, or the eight gluons of chromodynamics can be regarded as arising as a result of the local gauge symmetry of SU(3)). If the symmetry group, G, is not a direct product of subgroups, we can declare that T is fully unified; if G is a direct product of subgroups, T lacks full unity; and if the N entities are such that they cannot be regarded as arising as a result of some symmetry of T, with some group structure G, then T is disunified.[9] Example: T postulates the classical electromagnetic

---

8. Counting entities is rendered a little less ambiguous if a system of M particles is counted as a (somewhat peculiar) field. This means that M particles all of the same kind (i.e. with the same dynamic properties) is counted as *one* entity. In the text I continue to adopt the convention that M particles all the same dynamically represents one *kind* of entity, rather than one entity.

9. An informal sketch of these matters is given in (Maxwell, 1998, ch. 4, sections 11 to 13, and the appendix). For rather more detailed accounts of the locally gauge invariant structure of quantum field theories see: (Moriyasu, 1983; Aitchison and Hey, 1982, part III; and Griffiths, 1987, ch. 11). For a non-technical discussion of the role of symmetry and group theory in physics, see (Maxwell, 1998, pp. 257-265); for somewhat more

field, composed of the electric and magnetic fields, obeying Maxwell's equations for the field in the vacuum. The symmetry of Lorentz invariance unifies these two fields (see below). Here, N = 1 (in a type (6) way).

(7) If (apparent) disunity of there being N distinct kinds of particle or distinct fields has emerged as a result of cosmic spontaneous symmetry-breaking events, there being manifest unity before these occurred, then the relevant theory, T, is unified.   If current (apparent) disunity has not emerged from unity in this way, as a result of spontaneous symmetry-breaking, then the relevant theory, T, is disunified. Example: Weinberg's and Salam's electroweak theory, according to which at very high energies, such as those that existed soon after the big bang, the electroweak force has the form of two forces, one with three associated massless particles, two charged, W^- and W^+, and one neutral, W^o, and the other with one neutral massless particle, V^o.   According to the theory, the two neutral particles, W^o and V^o, are intermingled in two different ways, to form two new, neutral particles, the photon, γ, and another neutral massless particle, Z^o.   As energy decreases, the W^+, W^- and Z^o particles acquire mass, due to the mechanism known as spontaneous symmetry-breaking (involving the hypothetical Higgs particle), while the photon, γ, retains its zero mass.   This theory unifies the weak and electromagnetic forces as a result of exhibiting the symmetry of local gauge invariance; this unification is only partial, however, because the symmetry group is a direct product of two groups, U(1) associated with V^o, and SU(2) associated with W^-, W^+ and W^o.[10]

(8) According to general relativity, Newton's force of gravitation is merely an aspect of the curvature of spacetime.   As a result of a change in our ideas about the nature of spacetime, so that its geometric properties become dynamic, a physical force disappears, or becomes unified with spacetime.   This suggests the following requirement for unity: spacetime on the one hand, and physical particles-and-forces on the other, must be unified into a single self-

---

technical introductory accounts of group theory as it arises in physics see (Isham, 1989; or Jones, 1990).

   10. For accounts of spontaneous symmetry breaking see (Moriyasu, 1983; Mandl and Shaw, 1984; Griffiths, 1987, ch. 11).

interacting entity, U. If T postulates spacetime and physical "particles-and-forces" as two fundamentally distinct kinds of entity, then T is not unified in this respect. Example: One might imagine a version of string theory without strings, different vibrational modes (perhaps) of empty, compactified six-dimensional space giving rise to the appearance of particles and forces, even though in reality there is only 10 dimensional spacetime. Or one might imagine that the quantization of spacetime leads to the appearance of particles and forces as only apparently distinct from empty spacetime. In either case, $N = 1$: there is just the one self-interacting entity, empty spacetime.

For unity, in each case, we require $N = 1$. As we go from (1) to (5), the requirements for unity are intended to be cumulative: each presupposes that $N = 1$ for previous requirements. As far as (6) and (7) are concerned, if there are N distinct kinds of entity which are not unified by a symmetry, whether broken or not, then the degree of disunity is the same as that for (4) and (5), depending on whether there are N distinct forces, or one force but N distinct kinds of entity between which the force acts.

(8) does not introduce a new kind of unity, but introduces, rather, a new, more severe way of counting different kinds of entity. (1) to (7) require, for unity, that there is one kind of self-interacting physical entity evolving in a distinct spacetime, the way this entity evolves being specified, of course, by a consistent physical theory. According to (1) to (7), even though there are, in a sense, two kinds of entity, matter (or particles-and-forces) on the one hand, and spacetime on the other, nevertheless $N = 1$. According to (8), this would yield $N = 2$. For $N = 1$, (8) one basic entity (unified by means of a spontaneously broken symmetry, perhaps).

As we go from (1) to (8), then, requirements for unity become increasingly demanding, with (6) and (7) being at least as demanding as (4) and (5), as explained above.[11]

---

11. The account of theoretical unity given here simplifies the account given in (Maxwell, 1998, chs. 3 and 4), where unity is explicated as "exemplifying physicalism", where physicalism is the metaphysical thesis asserting that the universe has some kind of unified dynamic structure. Explicating unity in that way invites the charge of circularity, a charge that is not actually valid (see Maxwell, 1998, pp. 118-23 and 168-72). The

(1) to (8) may seem very different requirements for unity. In fact they all exemplify the same basic idea: disunity arises when *different* dynamical laws govern the evolution of physical states in different regions of the space, S, of all possible physical states. For example, if a theory postulates more than one force, or kind of particle, not unified by symmetry, then in different regions of S different force laws will operate. If different fields, or different kinds of particle, are unified by a symmetry, then a symmetry transformation may seem to change the relative strengths of the fields, or change one kind of particle into another kind, but will actually leave the dynamics unaffected. In this case, then, the same fields, or the same kind of particle are, in effect, present everywhere throughout S. If (8) is not satisfied, there is a region of S where only empty space exists, the laws being merely those which specify the nature of empty space or spacetime. The eight distinct facets of unity, (1) to (8) arise, as I have said, because of the eight *different* ways in which content can vary from one region of S to another, some differences being *more* different than others.[12] Some of the above requirements for unity are suggested by developments in 20th century physics. This is true in particular of (6) to (8). The important point, however, is that all these requirements for unity, (1) to (8), exemplify the same basic idea: a theory, in order to be unified, must assert that the same laws apply throughout the phenomena to which it applies. (1) to (8) in effect represent different, increasingly subtle ways in which a theory can fail to be unified in this sense, granted that $N > 1$ in each case.

One point demonstrated by (1) to (8) is that there is no sharp distinction between the very crudest kind of unity or non-*ad hoc*-ness, specified by (1) or (2), and the most demanding kind of theoretical unity, indicated by (8). One might, perhaps, draw a line between (1) to (3) on the one hand, and (4) to (8) on the other hand, and declare (1) to (3) to

---

account given here forestalls this charge from the outset.

12. The eight facets of unity, (1) to (8), specify eight different ways in which the content of a theory may fail to be the same throughout S (the space of states to which the theory applies) or may appear to fail to be the same when this is not really the case. This means that (1) to (8) are integral to physicalism, although of course progressively weaker versions of physicalism can be distinguished, beginning with all eight facets being assumed, then (1) to (5), (6) or (7), then (1) to (4), then (1) to (3), and so on. This idea is expounded and exploited in (Maxwell, 2004b), where an improved version of AOE is formulated.

be essential to physics, and (4) to (8) as being desirable but optional: but any such distinction is more or less arbitrary.[13]

Symmetry plays a role in almost all of the above eight facets of unity. What this role is, already indicated above, can be illustrated further by means of the example of classical electrodynamics and the symmetry of special relativity: Lorentz invariance. One might be inclined to hold that the electromagnetic field is made up of two distinct entities: the electric and the magnetic fields. But, as a result of being Lorentz invariant, the way the electromagnetic field divides up into the electric and magnetic fields depends on choice of reference frame. And, according to special relativity, nothing of fundamental dynamical significance can depend on choice of (inertial) reference frame. Any specific way we might choose to divide the field into the electric and magnetic fields would be arbitrary; it would be equivalent to choosing, arbitrarily, one frame, one state of motion to constitute absolute rest, from the infinitely many equally available, and all equivalent, according to special relativity. This would violate special relativity. Thus special relativity (Lorentz invariance) requires that we regard the electromagnetic field as *one* entity with two aspects, and not as *two* distinct entities.

To assert of a physical theory that it exhibits such and such a symmetry (invariance with respect to spatial position, time of occurrence, orientation, uniform velocity, and so on) is to assert, roughly, that, given any isolated physical system evolving in accordance with the theory, changing the system in some specified way — e.g. changing its location, orientation, time of occurrence, and so on — leaves the way the system evolves unaffected, according to the theory in question. If a system is moved some distance in space, is permitted to

---

13. This point is of fundamental importance for the problem of induction. Traditionally, the problem is interpreted as the problem of justifying exclusion of empirically successful theories that are *ad hoc* in sense (1): How can evidence from the past provide grounds for any belief about the future? This makes the problem seem highly "philosophical", remote from any problem realistically encountered in scientific practice. But the moment it is appreciated that the problem of justifying exclusion of empirically successful theories that are *ad hoc* in sense (1) is just an extreme, special case of the more general problem of excluding empirically successful theories that are *ad hoc* in senses (1) to (8), it becomes clear that this latter problem is a scientific problem, a problem of theoretical physics itself. For the implications of this crucial insight, and for a proposal as to how the problem of induction is to be solved exploiting it, see (Maxwell, 1998, especially chs. 4 and 5), and section 6 below of the present chapter.

evolve for some time, t, and is then moved back (instantaneously) to its original position, its state will be exactly what it would have been had it not been moved at all. The symmetry operation (of moving in space, changing orientation, and so on) must, in other words, commute with what may be called the "time-evolution operator" (associated with the theory in question), the operator that transforms the state of a physical system, from its state at one time to its state at another time.

A physical symmetry, P, is:

(a) A one-one mapping of the space of isolated systems into itself such that these mappings split this space into infinitely many equivalence classes, any point in any such equivalence class only being mapped into a point in the same class. These mappings form a group, G, the symmetry group of P.

(b) The one-one mappings, elements of G, must correspond to some physical operation, such as translation in space or time, change of orientation or velocity which leaves the way the systems evolve unaffected.

(c) These operations, O, (elements of G) must commute with the time-evolution operator, t, so that $O.t = t.O$.

For further clarification see (Maxwell, 1998, appendix).

As I have already in effect remarked, the current blueprint, B, at level 3 of Fig. 1.2, specifies, more restrictively than physicalism at level 4, what kinds of physical entity the world is composed of, what kinds of physical symmetry the theory of everything exhibits. Because B is much more specific and restrictive than physicalism, it is much more likely to be false. And the history of physics reveals that, again and again, versions of B have been put forward which have subsequently turned out to be false, as we have seen.

The explication of unity of theory given above serves also to explicate what it means to assert that the universe is physically comprehensible (i.e. physicalism, the level 4 thesis of AOE). For we can stipulate that to assert that the universe is physically comprehensible is to assert that there is a yet-to-be-discovered true physical theory of everything, and it is unified.

I now consider a number of objections to the account of the unity of a physical theory just expounded.

It may be objected that we never encounter the naked content of a theory, formulation free; we only encounter theories given some formulation. How, then, can we judge whether the *content* does or does not vary through the space S? The answer is that theories are not natural objects we stumble across; *we* formulate theories, and it is for us to ensure, granted we want our theories to be unified, that the content does not change as we move through S. We can arrange, however, that formulation matches content by ensuring that the terminology, the concepts, we use to formulate a theory do not surreptitiously change as we move through S. Given invariant concepts, if the form of the theory is also invariant throughout S, its content will be too. But if, for example, we surreptitiously change our units of length or mass as we move through space, then a theory whose content is spatially invariant will change its form with changes of spatial position, to the extent of an apparent change in the value constants (a point which will be taken up again below).

It may be objected that, given any theory, however unified, special regularities will always arise in restricted regions of S, which means disunity. Whether or not the theory is unified is, at best, ambiguous. Thus, given NT, in some regions of S there will be solar systems with planets that rotate in the same direction and conform to Bode's law, whereas in other regions of S these regularities or "laws" will be violated. The answer is to distinguish sharply between accidental and law-like regularities; only the latter are relevant for the assessment of unity. But how is this distinction to be made? I have solved this problem elsewhere: see (Maxwell 1968; 1998, pp. 141-155). Given a true law-like statement, this is a genuine physical law (having nomic necessity) if and only if physical dynamical (or necessitating) properties exist corresponding to the law. Thus Newton's law of gravitation can be interpreted as attributing the dynamical, necessitating property of Newtonian gravitational charge to massive objects. Objects that have this property of necessity obey Newton's law of gravitation. (The empirical content of NT, on this interpretation, is concentrated in the factual assertion: all massive objects possess Newtonian gravitational charge equal to their mass.) If no such property corresponds to a true regularity, then it is merely a true accidental regularity, and not a true law. (What is a necessitating property? This is explicated in Maxwell, 1968; 1998, pp. 141-155.) For unity we require that dynamical necessitating properties remain the same throughout S; the regularities of (some) solar systems, mentioned above, are not relevant because these regularities are not law-like, and no dynamical property exists corresponding to them.[14]

It may be objected that physical systems which possess symmetries that are also symmetries of the theory which determines their evolution, may evolve in accordance with a simplified version of the theory. Thus systems consisting of two spheres equal in every way rotating in a fixed circle about their centre of mass obey a simplified version of the dynamical laws of NT. This means there are regions of S where the dynamical laws are especially simple, and thus different from other regions. Does this mean the theory is correspondingly disunified? The answer is No. We need, again, to consider dynamical properties corresponding to dynamical laws. In the example just considered, if NT is true (interpreted essentialistically) then the spheres in question possess gravitational charge just like all other massive objects. It is just that, in the case of the systems possessing some rotational symmetry, the full, rich implications of the dynamical property of gravitational charge is not made manifest.

It may be objected that we may not know whether two formulations of a theory are just that, two formulations with the same physical content, or two distinct theories with distinct contents. Heisenberg's and Schrödinger's distinct formulations of quantum theory might be an example. This is correct but beside the point. The terminological problem arises when we reformulate a given theory, T, in a variety of ways, some simple and unified, some horribly complex and disunified, but we do this in such a way as to ensure quite specifically that the different formulations have precisely the same content, make precisely the same assertions about the world, this being something that we can always do. The solution to the problem proposed above is not in any way undermined by the fact that it sometimes happens that we do not know whether two formulations of a theory have the same or different contents. Nor is the distinction between form and content undermined: form has to do with what we write down on paper, content with what is being asserted. That we sometimes do not know whether difference of formulation ensures difference of content does not in the least

---

14. I am grateful to Jos Uffink for drawing my attention to the two objections just discussed.

undermine the distinction between form and content. It deserves to be noted, in addition, that one and the same formulation of a theory may be interpreted in more than one way, and thus may have different contents associated with it — a point which, again, does not undermine the theory presented here.

It may be objected the distinction between dynamical laws which do, and do not, remain the same throughout the space S cannot be maintained. Consider the two functions:

(1) y = 3x for all x
(2) y = 3x for x ≤ 2 and y = 4x for x > 2.

It is tempting to say that (1) remains the same as x changes, but (2) does not, since what (2) asserts changes at x = 2. But given the mathematical notion of function as a rule, (2) is just as good a function as (1) and, like (1), "remains the same" as x varies. Functions corresponding to physical theories are somewhat more elaborate than this, but the above point is not affected by that consideration: it seems that the very distinction between "remains the same" and "changes" as one moves through S collapses. Clearly, in order to meet this objection, functions corresponding to physical theories need to be restricted to a narrower notion of function than the above standard mathematical one, if we are to be able to distinguish between functional relationships which do, and which do not, "remain the same" as values of variables change. We need to appeal to what may be called "invariant functions", functions which specify some fixed set of mathematical operations to be performed on "x" (or its equivalent) to obtain "y" (or its equivalent). In the example just given, (1) is invariant, but (2) is not. (2) is made up of two truncated invariant functions, stuck together at x = 2. Functions that appear in theoretical physics are *analytic* (i.e. can be represented by a power series). Such functions have the remarkable property that from any small bit of the function, the whole function can be reconstructed uniquely, by a process called "analytic continuation". All analytic functions are thus invariant. The latter notion is however a wider one, and theoretical physics might, one day, need to employ this wider notion explicitly, if space and time turn out to be discontinuous, and analytic functions have to be abandoned at a fundamental level.

A similar remark needs to be made about Goodman's (1954) paradox concerning "grue" and "bleen". Modifying the paradox slightly, an object is grue if it is green up to time t, blue after t; it is bleen if it is blue up to time t, green after t. Sometimes it is held that there is perfect symmetry between blue and green, on the one hand, and grue and bleen on the other, especially as "emeralds are green" is equivalent to "emeralds are grue up to t, and bleen afterwards". But this symmetry is merely terminological and, as we have already seen in connection with the *ad hoc* version of NT discussed above, terminological symmetry does not mean there is symmetry of content. That there is not symmetry of content in the grue/bleen case can be demonstrated as follows. If emeralds are grue, a person convinced of this can determine whether time t is future or past merely by looking at emeralds. But if emeralds are green, a person convinced of this cannot say whether t is in the future or past by just looking at emeralds. The content, the meaning, of grue and bleen contains an implicit reference to t in a way in which that of green and blue do not. Doubtless symmetry can be created by considering two possible worlds, ours and a Goodmanesque one with special physics and/or physiology of vision so that grue emeralds do not appear to change at t, whereas green ones do. This, however, is to consider conditions quite different from those specified by Goodman. The crucial point to make, in any case, is that dynamical or physical properties, of the kind attributed to physical entities by physical theories (interpreted in a conjecturally essentialistic way), are like blue and green, and unlike grue and bleen, in not containing any implicit reference to specific times or places (or to hypersurfaces of S that distinguish one region of S from another). Physical properties must be *invariant* in a sense that corresponds to the *invariance* of allowed functions in physics. The more general notion of property, which includes Goodmanesque properties, is excluded, just as the more general notion of function, which includes (2) above as an "unchanging" function, is excluded.

It deserves to be noted that the above account of unity provides us with a second, quite distinct methodologically significant feature of theories which is distinct from unity, and which we may call "simplicity". The simplicity of a theory can be interpreted as having to do, not with whether the *same* laws apply throughout the space S, but rather with the *nature* of the laws, granted that they are the same. Some laws are simpler than others. The problem, here, is not to say what "simplicity" means (it should be understood in its ordinary sense) but to solve the problem that the simplicity of a theory would seem to be highly dependent on its formulation (Maxwell, 1998, pp. 157-158). In order to solve this problem it is essential, as in the case of unity, to interpret "simplicity" as applying to the *content* of theories, and not to their *formulation*, their *axiomatic structure*, etc. Theories can only, at best, be partially ordered with respect to degrees of simplicity, as in the case of unity. Even when two theories are amenable to being assessed with respect to relative simplicity, there is always the problem that a change of variables may reverse the assessment. Let the two theories be (1) $y = x$, and (2) $y = x^2$. We judge (1) to be simpler than (2). Let $x^2 = z$. We now have (1) $y = \sqrt{z}$, and (2) $y = z$. Now (2) is simpler than (1). Assessment of relative simplicity of two theories may only be unambiguous when restrictions are placed on the form that physical variables can take, so that only linear transformations of the type $z = Ax + B$ (where A and B are constants) are permitted for example. It is a further great success of the theory presented here that it succeeds in distinguishing sharply between these two aspects of physical theory, the *unity* and *simplicity* aspects, and succeeds in explicating both.[15]

These two notions can be used to solve the problem of ambiguity of judgement concerning the relative non-empirical merits of Newton's and Einstein's theories of gravitation. Einstein's theory is more *unified* in that it eliminates gravitation as a force distinct from space and time, and reformulates Newton's first law so that it becomes the assertion that bodies move along geodesics in curved space-time, curvature being caused by mass, or by stress-energy-density more generally. On the other hand Newton's theory is *simpler*, in that Einstein's field equations are really six independent equations which, taken together, are more complex than the single equation which determines the Newtonian gravitational field (Schutz, 1989, pp. 195-200). For the purposes of this comparison one must attend to the *content* of these equations, what they assert about physical reality, and not merely to the *form*, which could vary with changes of formulation. Incidentally, it is reasonable to expect that, as theoretical physics draws closer to capturing the true theory of everything, the totality of fundamental physical theory will become increasingly unified, and at the same time increasingly complex. Unity

---

15. For further discussion of simplicity, and how terminological simplicity can be related to unity, see (Maxwell, 1998, pp. 110-113 and 157-159).

outweighs simplicity in methodological importance. It is essential for acceptable physical theories to be (more or less) unified, desirable for them to be simple. (Simplicity is methodologically significant if it repeatedly happens that a unified theory, T, is the simplest of a family of similar, equally unified theories, and T turns out to be acceptable on empirical grounds, and the other theories turn out eventually to be unacceptable — or would so turn out if empirically investigated.)

3 Arguments For and Against Aim-Oriented Empiricism (AOE)

Once it is conceded that scientific method commits science to making some kind of metaphysical assumption concerning an underlying dynamic unity in nature, the crucial question becomes: What ought this assumption to be? What is the rational choice? In particular, what grounds are there for choosing the level 4 thesis of AOE of physicalism, the thesis that the universe is perfectly physically comprehensible — such that the true physical theory of everything is perfectly unified and explanatory (in so far as this is logically possible)?

Rivals to AOE that consist of just one metaphysical thesis, and not a hierarchy of theses, fail miserably when compared to AOE. Such views fail, either because (i) they are much too specific for their acceptance to be justified on the grounds that, if false, the acquisition of knowledge becomes impossible, or because (ii) they are much too unspecific to exclude empirically successful but grotesquely disunified theories, or because (iii) of both (i) and (ii). Such a rival might consist just of the level 4 thesis of AOE of physicalism. This does not fall to objection (ii) because it is sufficiently specific to exclude grotesquely disunified theories; but it does fall to objection (i) in that knowledge could be acquired in a universe in which physicalism is false. Or the rival might consist of the thesis at level 7 of AOE (see below). Such a view does not fall to objection (i), because if the thesis is false then acquisition of knowledge does become impossible; it does, however, fall to objection (ii), in that there will always be infinitely many empirically successful but grotesquely disunified theories compatible with level 7. Views which appeal to metaphysical theses somewhere between the generality of level 7 and the specificness of level 4 will fall to *both* (i) and (ii).

A serious rival to AOE must, then, take the form of postulating a hierarchy of metaphysical theses. Let us call any such view a version of "generalized AOE" (GAOE). I propose now to consider what reasons there are for accepting AOE in preference to rival versions of GAOE. But before I do this, I must first state, a little more explicitly, the various theses asserted by AOE at levels 1 to 7: see Fig. 1.2.

Level 1: Empirical phenomena.

Level 2: Accepted testable fundamental physical theories (or laws governing phenomena in the absence of a theory); at present general relativity plus the so-called standard model.

Level 3: Best blueprint — a specific version of physicalism. The best available more or less specific metaphysical view as to how the universe is physically comprehensible, a view which asserts that everything is composed of some more or less specific kind of physical entity, all change and diversity being, in principle, explicable in terms of this kind of entity. Historically important versions of this thesis include: the 17th century idea that the universe is made up of rigid corpuscles that interact by contact; the 18th and early 19th century idea that the universe is made up of point-atoms, each surrounded by a rigid, spherically symmetrical force field, alternatively repulsive and attractive; the late 19th and early 20th century idea, associated especially with Faraday and Einstein, that the universe is made up of a unified self-interacting classical field; the idea that it is made up of some kind of quantum field; the idea that it is made up of empty spacetime with variable curvature and possibly also variable topology; the idea that this is quantized; the late 20th and early 21st century idea that the universe is made up of quantum strings in ten or eleven dimensions of spacetime.

Level 4: Physicalism — the thesis that the universe is physically comprehensible, there being some kind of unified physical entity (or kind of entity) that exists at all times and places, throughout all phenomena, which determines (perhaps probabilistically) the way events unfold, and in terms of which all physical phenomena could (in principle) be explained and understood. If physicalism (in the strongest sense) is true, then the true theory of everything satisfies all eight facets of unity.

Level 5: Comprehensibility — the thesis the universe is comprehensible in some way, there being a perfect, overall explanation (of some kind of other) for all the diverse phenomena of the universe (in so far as this is logically possible). What this amounts to is that there is *something* — God, society of gods, overall cosmic purpose, cosmic programme, or physical entity — which is present at all times and places, throughout all phenomena, which determines (perhaps probabilistically) the way events unfold, and which can thus provide an explanation as to why events occur as they do.

Level 6: Meta-knowability — the thesis that the universe is such that there is some true, rationally discoverable thesis which, if accepted, enables us progressively to improve more specific assumptions and methods in the light of empirical success and failure, thus progressively improving methods for the improvement of knowledge.

Level 7: Partial knowability — the thesis that the universe is such that we can acquire some knowledge of our local circumstances, sufficient at least to make life possible.

This depiction of knowledge in physics is of course highly schematic, and comes hedged about with all sorts of qualifications (see Maxwell, 1998, pp. 32-33). Earlier versions of AOE (see for example, Maxwell, 1998, pp. 7-13) included three additional theses, two of which are designed to take into account the possibility that the universe is only approximately or partially comprehensible. I now feel these theses are redundant. AOE is a highly specific version of GAOE which does the best justice (I will argue) to science as it exists at present: it is conceivable that it will need drastic revision in the future. This possibility is provided for by arranging for AOE to exist within the broader framework of GAOE.

Despite asserting only that our local circumstances are such that we can continue to acquire knowledge sufficient to make life possible, partial knowability is nevertheless a cosmological thesis, with implications for the entire universe. It implies that conditions throughout the universe are such as to permit us to continue to acquire knowledge locally. It deserves to be noted that even our most trivial, common sense items of knowledge, such as my knowledge that the room I am in at the moment will continue to exist undisturbed for the next ten minutes, contains a cosmological dimension. It implies that there is no vast cosmic explosion occurring billions of light-years away which will travel at near infinite speeds to engulf and destroy the room before the ten minutes are up. Thus we only have such trivial common sense items of knowledge in so far as we have some knowledge about the entire cosmos. Once this point has been noted, my claim that scientific knowledge contains a cosmological dimension should seem much more acceptable.

The level 6 thesis of meta-knowability asserts, in effect, that the universe is such that we can improve our knowledge by putting the meta-methodology of GAOE into practice. It asserts that there is *some* true, rationally discoverable thesis which is such that, if accepted, it would form the basis for the successful implementation of the corresponding version of GAOE. Granted that we are rationally entitled to accept meta-knowability, then we are rationally entitled to implement GAOE, but this does not mean, of course, that any version of GAOE we happen to hit upon will meet with real success. ("Some" does not mean "any".) "Rationally discoverable" means that the thesis is implicit in our existing culture and can be made explicit; it means, at least, that the thesis does not represent an arbitrary choice from infinitely many analogous theses. Einstein endorsed a version of this last requirement in this way: "The following I reckon as also belonging to the "inner perfection" of a theory: We prize a theory more highly if, from the logical standpoint, it is not the result of an arbitrary choice among theories which, among themselves, are of equal value and analogously constructed" (Einstein, 1949, p. 23).

The level 5 thesis of comprehensibility is, I claim, rationally discoverable in the way I have indicated. But consider the thesis that the universe is such that the true physical theory of everything is disunified in some specific way in one or other of the senses (1) to (3). This thesis is not rationally discoverable as it does represent an arbitrary choice from infinitely many analogous theses — there being infinitely many rival theories of everything just as disunified as the chosen one.

What matters, of course, is that a thesis is "rationally discoverable" relative to *our* culture. The fact that a thesis is "rationally discoverable" relative to some crazy, alien culture, with crazy ideas about unity and sameness, is neither here nor there. (This point is relevant to the Goodman paradox, briefly discussed above.)

AOE is Popperian in character, in that it accepts that all knowledge is conjectural, and that the fundamental problem of epistemology and methodology is: how can we best go about improving our conjectural knowledge? It agrees with Popper in holding that scepticism, far from being the enemy of reason, is close to being the heart of reason. The critical attitude is vital for the improvement of knowledge. AOE disagrees with Popper, however, in recognizing that criticism is only rational in so far as it contributes to the improvement of knowledge. As I shall stress below when discussing the problem of induction, AOE fine tunes Popperianism in recognizing that we need to be *critically* critical. For example, some theses are such that criticism of them cannot, in any circumstances, contribute to the improvement of knowledge because, if false, the acquisition of knowledge becomes impossible: these do not require sustained critical scrutiny. Again, among other theses accepted as a part of knowledge some, we may have grounds to conjecture, are more likely to be fruitful to criticize than others. A vital task for the philosophy of science is to discriminate between theses on these grounds and, indeed, to grade theses in terms of how fruitful they are to criticize from the standpoint of the improvement of knowledge. It is just this which the hierarchy of AOE does. The two considerations just indicated (which fine tune Popperianism) lead to the hierarchy of metaphysical theses of AOE, so dramatically at odds with Popper's falsificationist version of standard empiricism (SE).

AOE is more rigorous than any version of SE because, unlike SE, it puts into practice the following elementary requirement for intellectual rigour, thoroughly Popperian in spirit (stated in section 1.3 of chapter one):

Principle of Intellectual Integrity: Assumptions that are substantial, influential, problematic and implicit need to be made explicit so that they can be critically assessed, so that alternatives can be developed and considered, in the hope that this may lead to the assumptions being improved.

Persistent preference for unifying, explanatory theories in science, even against the evidence, means that science assumes (perhaps implicitly) that the universe has a unified dynamic structure, to some degree at least. Science is more rigorous if this implicit assumption is made explicit rather than denied because, as a result of being acknowledged explicitly, the assumption becomes more readily criticizable, and thus open to improvement. The end product of further enhancing the rigour of science by making explicit what most needs to be improved, and thus subjected to criticism, is AOE. AOE improves on its rivals because it facilitates the growth of knowledge better than any of its rivals by subjecting to critical scrutiny what most needs to be so subjected, for the growth of knowledge: it is, in this way, more rigorous than its rivals.

(Rigour is often thought of, especially in mathematics, in such a way that it seems to obstruct, rather than facilitate, progress; but this just indicates that rigour is being misconstrued in justificationist terms, as being concerned exclusively with decisive proof, establishing certainty, removing doubt. Within science, however, rigour, properly construed, has to do with promoting the growth or, better, the improvement,[16] of knowledge. It is a crucial part of scientific rigour, then, to try to ensure that what stands in most need of improvement is subjected to the fiercest critical scrutiny. The greater rigour of AOE over any rival has to do with the fact that AOE, better than any rival, arranges for that which most needs critical scrutiny to receive just this treatment. Effective criticism depends, of course, on derivations from the thesis under critical scrutiny being of the utmost reliability. The requirement that scientific rigour requires searching criticism of what most needs criticism in turn demands that logic and mathematics are rigorous in the rather different sense that derivations are as secure, as reliable, as possible. It may be, of course, that the best way to secure derivational reliability is to subject ostensible derivations to severe criticism. And it may be, as Lakatos has suggested, that a "proof" increases the vulnerability of the theorem in question to refutation, in that the "proof" aids the discovery of potential counter-examples: see (Lakatos, 1976). Proof increases security by facilitating the discovery of counter-examples if they exist.)

No attempt is made, then, to argue that any of the theses of AOE stand a better chance of being *true* than any rival theses.[17] Instead, it is argued that, granted that our aim is to improve knowledge of truth (in so far as this is possible), then we are justified in *accepting* the theses of AOE as a part of knowledge in preference to any rival theses. Considerations appealed to include the following:

(a) P is to be accepted because if P is false, the acquisition of

---

16. Popper speaks of the "growth" of knowledge because, for him, empirical content is the decisive measure of progress. But according to AOE, science makes progress if explanations improve, perhaps because greater fundamental theoretical unity is achieved. This could come about without any increase in content. Thus "improvement" in knowledge seems more appropriate than merely the "growth" of knowledge.

17. This is not quite correct. I do argue that, as we go up the hierarchy, theses are increasingly likely to be correct because they assert less and less.

knowledge becomes impossible.
(b) P is to be preferred to Q because P is potentially more fruitful for the growth of empirical knowledge.
(c) P is to be preferred to Q because P has actually been more fruitful for the apparent growth of empirical knowledge.
(d) P is to be preferred to Q because P is more readily critizable than Q.
(e) P is to be preferred to Q because it accords better with the accepted thesis next up in the hierarchy.[18]

Here, then, very quickly, is an indication of the kind of arguments to be employed in support of accepting the theses of AOE in preference to rival theses.

Partial knowability at level 7 is to be accepted because this can only help and cannot hinder the pursuit of knowledge; if the thesis is false, the acquisition of knowledge is impossible whatever is assumed.

Meta-knowability at level 6 is to be accepted for the following two reasons:

(i) Granted that there is *some* kind of general feature of the universe which makes it possible for us to acquire knowledge of our local environment (as guaranteed by the thesis at level 7), it is reasonable to suppose that we do not know all that there is to be known about what the *nature* of this general feature is. It is reasonable to suppose, in other words, that we can improve our knowledge about the nature of this general feature, thus improving methods for the improvement of knowledge. Not to suppose this is to assume, arrogantly, that we already know all that there is to be known about how to acquire new knowledge. Granted that learning is possible (as guaranteed by the level 7 thesis), it is reasonable to suppose that, as we learn more about the world, we will learn more about how to learn. Granted the level 7 thesis of partial knowability, in other words, meta-knowability is a reasonable conjecture.

---

18. How can P accord better that Q with R? This will be the case if P implies R and Q is only compatible with, or is incompatible with R. But even if P and Q are both incompatible with R, still P might accord better with R than Q, because R specifies or implies symmetry principles which Q violates in more ways than P does. As a result of specifying symmetry principles, R imposes a partial ordering on incompatible theses.

(ii) Meta-knowability is too good a possibility, from the standpoint of the growth of knowledge, not to be accepted initially, the idea only being reluctantly abandoned if all attempts at improving methods for the improvement of knowledge fail.

These two reasons for accepting meta-knowability are, admittedly, very weak. Nevertheless, granted that we can learn (guaranteed by level 7) it is hard not to conclude that we can also learn how to learn (or learn how to learn better), the nub of meta-knowability.

Comprehensibility, at level 5, is to be accepted because it is the best available candidate for acceptance, granted meta-knowability. It satisfies the meta-knowability criterion of being rationally discoverable. It is of great potential fruitfulness in that, if correct, it tells us that the diverse explanations we have of phenomena are diverse, imperfect versions of one overarching true explanation, there thus being a built-in heuristic for improving our existing explanations. And it is a thesis that has already seemed to be empirically fruitful, in that explanatory theories have met with empirical success. These points were valid even before the advent of modern science. The last point becomes overwhelmingly valid today, granted the immense (apparent) empirical success of modern science. There is no other thesis, comparable to comprehensibility in generality, that has been so fruitful empirically, in the sense indicated above (in the sense, that is, of leading to a research programme of immense empirical success - the research programme of modern natural science).

It is an easy matter to formulate *ad hoc* rivals to comprehensibility which assert that the universe is only partially or approximately comprehensible in various ways,[19] each of which can be construed to support an *ad hoc* research programme, with *ad hoc* versions of accepted scientific theories, that is just as empirically successful as modern science interpreted in terms of AOE. But all such theses are ruled out because they violate meta-knowability in failing to be "rationally discoverable". Any such assumption of partial comprehensibility

---

19. See (Maxwell, 1998, pp. 168-172) for a long list of different kinds of assumption concerning partially comprehensible universes, each capable of supporting *ad hoc* research programmes just as empirically successful as modern science, there being infinitely many distinct kinds of partially comprehensible possible universes corresponding to each item on the list.

constitutes an arbitrary choice from infinitely many equivalent assumptions. Such an assumption violates Einstein's requirement, indicated above.

Physicalism at level 4 is to be accepted because no other version of the level 5 thesis of comprehensibility is remotely as empirically fruitful as this thesis is. As I have already argued, all theoretical revolutions in physics have exemplified the persisting idea of unification, from Galileo and Newton, via Faraday, Maxwell, Einstein, Bohr, Heisenberg, Schrödinger, Dirac, Tomonaga, Schwinger, Feynman and Dyson, to Yang, Mills, Gell-Mann, Weinberg and Salam.[20]

Choosing the best "blueprint" at level 3 is both difficult and enormously important for theoretical physics. What is required is a specific version of physicalism which unifies, at a metaphysical level, the fundamental ideas of the so-called standard model (the quantum theory of fundamental particles and fields) and general relativity. This might be some kind of probabilistic dynamic spacetime geometry. Perhaps locally there are quasi quantum mechanical superpositions of spacetimes of different curvatures which collapse probabilistically into just one such curved spacetime on space-like hypersurfaces.[21] Perhaps some development of the ideas of string theory would provide a good blueprint. Or perhaps something much more radical is required which postulates a discrete spacetime. I propose a much more conservative blueprint: Lagrangianism. This asserts that the universe is such that all phenomena occur in accordance with Hamilton's principle of least action, formulated in terms of some unified Lagrangian (or Lagrangian density), L. L must not be the sum of two or more distinct Lagrangians; it must have a single physical interpretation, and its symmetries must have an appropriate group structure.

In support of Lagrangianism one can say that all current physical theories can be given a Lagrangian formulation, it being thus not entirely implausible to suppose that the true unified theory of everything would also have this form, and would have the unified character just indicated.[22]

What rival versions of GAOE deserve consideration? I now consider,

---

20. For a more detailed argument along these lines see (Maxwell, 1998, ch. 5).

21. I have put this suggestion forward in (Maxwell, 1985b).

22. Lagrangianism refers, of course, to the *content* of the theory of everything, not to its *form* given this or that formulation.

very briefly, one or two possible rivals to theses of AOE, going up from level 3, and thus differing from AOE in increasingly substantial ways.

It is very likely that Lagrangianism is false. Recent work on quantum gravity, having to do with duality, suggests this: see, for example, (Isham, 1997, pp. 194-195). If spacetime is discrete then Lagrangianism is false. Since Galileo's time, ideas at the level 3 blueprint level have changed many times, as I have already indicated; this fact alone suggests that our current best idea at this level is unlikely to be correct. A new metaphysical idea, compatible with physicalism, which succeeds in unifying the basic ideas of SM and GR, would be an important contribution to theoretical physics.

What about rivals to physicalism? One obvious candidate is the thesis that the universe is comprehensible because God created it that way. There are two versions of this view. The first holds that all physical phenomena conform to physicalism, while the second holds that some phenomena violate physicalism (because God wills it). Neither is as acceptable as physicalism *per se*. The first is more *ad hoc* than physicalism *per se*, in that God is postulated but plays no explanatory role. The second only promises to be as empirically fruitful as physicalism *per se* if the nature of God (His goals, plans, *modus operandi*) is specified in a sufficiently definite way to generate a research programme more empirically successful than that of science. As long as the nature of God is unknown and unknowable, this version of the God thesis remains less acceptable than physicalism *per se*, because it holds out less hope of being empirically fruitful, and because, where it differs from physicalism *per se*, it has no apparent empirical fruitfulness at all.[23]

In order to be at least as acceptable as physicalism, all rival level 4 theses that retain comprehensibility at level 5 must be both (a) compatible with comprehensibility, and (b) as empirically fruitful as physicalism. It is not easy to think up an idea that fulfils both (a) and (b), but below, in section 5, I consider a possible candidate.

What about rivals to the level 5 thesis of comprehensibility? It is important for AOE that such rivals are possible, for if they are not, and accepting meta-knowability at level 6 commits one to accepting comprehensibility at level 5, then meta-knowability becomes redundant,

---

23. For further criticisms of this historically important rival to physicalism see (Maxwell, 1998, pp. 206-208; 2001c, pp. 6-10).

in that it implies comprehensibility. Fortunately it is easy to think of rivals to comprehensibility that exemplify meta-knowability. One such rival is the idea that the universe is such that, in order to advance knowledge, a succession of revolutions is required, the number of unified theories required to cover all phenomena at the given level going up by one. Another rival would be the thesis that there are three distinct forces in nature; each force is such that, in order to capture its character precisely, infinitely many theoretical revolutions are required. A third rival might be called asymptotic comprehensibility: the universe is such that infinitely many theoretical revolutions are required to capture its unified dynamic structure precisely.

Comprehensibility is to be preferred to all such rivals because of its greater heuristic power, its greater potential and (apparently) actual empirical fruitfulness.

What about rivals to meta-knowability? This thesis might be rejected altogether. Or, alternatively, a much weaker version of the thesis might be accepted, which does not require that theses accepted lower down in the hierarchy are "rationally discoverable" in the sense indicated above. The two arguments given above, (i) and (ii), do give reasons for preferring meta-knowability to such rivals, but these reasons are, it must be admitted, weak. An additional argument in support of accepting meta-knowability is this. A version of GAOE that accepts such a rival to meta-knowability at level 6 will, it seems be vulnerable to the four objections to LSE (Lakatosian AOE), indicated above.

One feature of AOE which many may find objectionable is that theses at levels 3 to 5 are incompatible with current fundamental theoretical scientific knowledge. Accepted fundamental theory in physics at present consists of two very different theories, the so-called standard model (SM), the locally gauge invariant quantum field theory of fundamental particles and forces, and general relativity (GR). This conjunction, SM + GR, is incompatible with the thesis that the universe is perfectly comprehensible, demanded by levels 3 to 5 of AOE. Surely scientific knowledge, even if theoretical, must take precedence over mere metaphysical speculation!

But this ignores, in the first place, that physicalism (or comprehensibility) is not merely idle metaphysical speculation. It is implicit in the whole enterprise of theoretical physics, from Galileo onwards, or from Thales onwards over two and a half thousand years ago (with a few wobbles to take in such things as Aristotelianism and Christianity). Natural science searches single-mindedly for explanations for phenomena, the whole enterprise presupposing that these exist to be discovered, the very term "theory" having the idea of explanation built into it, so that "non-explanatory theory" is almost a contradiction in terms. The methods of physics, pushed to the limit, presuppose physicalism. All that AOE does, in accepting physicalism as a part of scientific knowledge, is to make explicit what is implicit in scientific method and practice.

Secondly, the above point ignores that new theories almost always contradict their precursors (as I have already remarked). Newtonian theory contradicts Kepler and Galileo; Maxwellian electrodynamics contradicts Newton; Einstein's special and general relativity contradict Newton; quantum theory contradicts classical physics; quantum field theory contradicts non-relativistic quantum theory. Given this past record, it is entirely reasonable to suppose that new and better fundamental physical theories to be discovered in the future will contradict present theories, this being the case, too, of the true theory of everything. This means that assumptions at levels 3 to 5 have a better chance of being true if they do contradict existing theoretical knowledge.

In the third place, versions of GAOE that accept theses at levels 3 to 5 that are compatible with existing theoretical knowledge do not look very plausible.

Consider, for example, "weak AOE" (WAOE).[24] This asserts that, at any stage in the development of physics, the strongest acceptable metaphysical thesis asserting unity is such that it is compatible with the totality of current fundamental physical theory — today, general relativity (GR) and the standard model (SM), consisting of quantum electroweak theory and quantum chromodynamics).[25] The moment one

---

24. Elsewhere I have formulated a wide range of versions of generalized AOE and argued that none is as good as AOE: see (Maxwell, 1998, pp. 223-227).

25. The standard model might be said to consist of three forces, electromagnetism, the weak force (associated with certain kinds of radioactive decay) and the strong force (a force between quarks which holds them together in threes to form protons and neutrons). Quantum electrodynamics is the quantum field theory of the electromagnetic force, and quantum chromodynamics is the quantum field theory of the strong force. There is also quantum electroweak theory, which is a quantum field theory that partially unifies the electromagnetic and weak forces. The standard model thus consists of the conjunction of the electroweak theory and chromodynamics. But because of the similar structure of these

tries to formulate the strongest thesis of WAOE all sorts of difficulties arise, and it looks thoroughly arbitrary and implausible. Consider the following formulation:

> (W) The universe is such that its degree of unity is no worse than the following. Spacetime is unified to degree $N = 1$ in type (1) to (7) ways (see section 2 above). Matter is unified to degree $N = 3$ in a type (4) way, in that there are three distinct kinds of forces (electromagnetism and the weak force being only partially unified); it is unified to degree $N = 24$ in a type (5) way, in that there are 24 distinct kinds of particle, with distinct properties, such as mass, not determined by theory.

But the situation is even worse than this, for a number of reasons. GR predicts the existence of black holes; inside each black hole, GR predicts, there is a singularity, a point of infinite curvature, which contradicts the idea that spacetime is everywhere smooth and differentiable. GR thus predicts its own downfall. Furthermore, SM predicts that quantum effects will manifest themselves before classical singularities emerge. GR and SM thus explicitly clash in such regions of high energy density. In the absence of a theory of quantum gravity, this can only be avoided by restricting the scope of the two theories in a highly *ad hoc* way, which drastically decreases the unity of GR + SM, in a type (2) way. And in addition SM, compounded of quantum field theories, suffers from the inherent defect of all versions of orthodox quantum theory (OQT), non-relativistic and relativistic. OQT is a theory about performing measurements on quantum systems, the measuring apparatus itself described by means of some other theory, usually some part of classical physics. This means, as we saw in chapter two, that the theory which makes unconditional predictions (and not just conditional predictions of the form "if such and such a measurement is made, the outcome will be such and such with such and such probability"), is an *ad hoc* composition of quantum postulates plus classical postulates.[26] SM is

---

two theories (they are both locally gauge invariant theories), the standard model is somewhat more unified than merely the arbitrary conjunction of these two theories. For further details see (Maxwell, 1998, pp. 125-14 0, ch. 7, and appendix), and additional literature referred to therein.

26. See Maxwell (1998, pp. 230-235) and additional references cited therein.

thus a severely *ad hoc* theory. There are additional unsatisfactory features of SM, due to the fact that the quantum field theories of which it is composed must be renormalized if they are to yield finite predictions. Renormalization is a somewhat dubious procedure that involves subtracting infinities from infinities to obtain finite predictions. Finally, there is dark matter, believed to account for a large percentage of mass of the universe, whose nature is at present unknown, and unaccounted for by present theory. The current theory of everything is thus GM + SM + ?, where "?" stands in for the unknown nature, and unknown theory, of dark matter.

If (W) is to be compatible with GM + SM + ?, it will need to be much weaker than the formulation of (W) given above.

Given the very messy, arbitrary character of thesis (W) of WAOE, it may be judged better to defend, as best rival to AOE, a doctrine which might be called "modest AOE" (MAOE). MAOE accepts, as its strongest metaphysical assumption, some combination of (1) to (8) of section 2 above which falls short of *all* of (1) to (8). For example, this thesis might be:

(M) The universe is such that the true physical theory of everything is unified to extent $N = 1$ in type (1) to (3) ways, and is unified to extent $N < 50$ in type (4) and (5) ways.

But (M) seems hardly less arbitrary than (W).

WAOE and MAOE are not just unsatisfactory because of the arbitrariness of the theses they incorporate. These theses fail to do justice to what theoretical physics presupposes. In attempting to discover better theories than those that we have at present physicists only consider theories that can be formulated in some modification of concepts employed to state existing accepted theories (SM and GR). This means theoretical physics makes a very much more precise persistent metaphysical assumption than either (W) or (M). (W) and (M) fail to do justice to the quite specific character of what theoretical physics actually presupposes. (This charge might be levelled against Lagrangianism as well. It depends on with what degree of specificness or generality Lagrangianism is formulated.)

The most decisive objection to WAOE and MAOE is that neither has any potential for being empirically fruitful, whereas AOE, by contrast, has great potential in this respect. This contrast can be brought out by making the simple observation that AOE provides physics with a fallible but rational method for the discovery of new fundamental physical theories, whereas neither WAOE nor MAOE can do any such thing.

In response to this, it might be argued that WAOE or MAOE could make use of the distinction between the context of discovery and justification. In the context of discovery, WAOE or MAOE accepts physicalism, and even Lagrangianism (or some more appropriate blueprint); but in the context of justification, these theses are not accepted, and thus are not a part of scientific knowledge. The problem with this view is that if these theses are not a part of knowledge, why should they be taken seriously when it comes to the task of discovering new fundamental physical theories?

A final point to note is this. AOE has the following great advantage over more modest competitors (such as WAOE and MAOE): its theses of comprehensibility at level 5, physicalism at level 4, and Lagrangianism at level 3, because of their greater substance and precision, are much more criticizable than rival more modest theses. If these theses are false, and require revision or replacement, it will be easier to discover that this is the case, than it would be if less substantial and precise rivals were accepted. AOE is more rigorous than rival views, as I have stressed above, because it is better at exposing what most needs to be exposed to criticism. It facilitates better learning about how to learn.

Another quite different rival to AOE that I need to consider is Bayesianism: for expositions see (Howson and Urbach, 1993; Howson, 2000); for a critical assessment see (Earman, 1992). Bayesianism (B) interprets probability as applying to propositions. The probability, p(H), assigned to a hypothesis, H, is to be interpreted as an assessment of the likelihood of H being true. Only those universal hypotheses assigned probabilities greater than zero prior to evidence can, according to B, receive support from subsequent evidence. If the prior probability of H is zero, its probability remains zero, however much evidence may be accumulated ostensibly in support of H.

Is B a version of SE? On the face of it, yes. The probabilistic apparatus of B is devoted to specifying how probabilities of (testable) theories are to be modified in the light of evidence. This accords perfectly with SE.

However, some proponents of B hold it to be quite different from other (SE) views. Thus Colin Howson, a leading proponent of B, remarks that it is an elementary feature of the view that "*evidential data cannot alone determine the credibility of a hypothesis*", and then goes on to say "This feature distinguishes the Bayesian account from practically every other model of scientific reasoning — to the detriment of the latter" (Howson, 2000, p. 180). It may seem from this that B, far from being a version of SE, stands with AOE in being in sharp contrast with "practically every other" SE view. But it is only *bare* SE that holds that evidence *alone* determines acceptability of theories. Most versions of SE (versions of *dressed* SE) accept that considerations such as simplicity, unity or explanatory power play a role *in addition* to evidence. In order to be distinct from SE, B must reject the central tenet of SE, namely "science makes no persistent assumption about the world independently of evidence". According to B, scientists legitimately assign probabilities to hypotheses greater than zero independently of evidence, but B does not assert that this involves committing science permanently to some metaphysical thesis about the world independently of evidence. B's failure explicitly to reject the central tenet of SE means, in effect, that B is a version of SE.

The more important question, however, is: do the reasons for rejecting all versions of SE, spelled out in chapter one, apply to B as well? The answer is yes. Like other versions of SE, B fails to make explicit, and so subject to critical scrutiny and improvement, metaphysical assumptions implicit in the way science persistently favours simple, unified theories (by assigning them non-zero prior probabilities in the case of B). B is a part of the neurotic face of science.

A number of other defects inherent in B, characteristic of all versions of SE, flow from the basic defect just indicated:

(1) It is crucial to the Bayesian account of science that appropriate prior probabilities are accorded to appropriate theories. Given any "accepted" physical theory, T, there will always be endlessly many disunified variants, $T_1, T_2, \ldots$, which will, on the face of it, receive all the empirical support that T receives, it being for a long time impossible (in practice or in principle) to perform experiments capable of verifying T and refuting $T_1, T_2, \ldots$. B can only eliminate $T_1, T_2, \ldots$ from science before these experiments are performed by

assigning zero prior probabilities to these theories (or at least prior probabilities very small compared to that assigned to T). But what is the basis for these assignments? Physicists are better at making such assignments than lawyers, let us say, or theologians or pilots, presumably because in doing physics they have learned something relevant about the world. What have they learned? Explicitly or implicitly, something like the hierarchy of metaphysical theses of AOE. But B cannot, of course, give this answer. At most, B can declare that physicists have learned to assign prior probabilities appropriately. But this does not amount to an objective factual feature about the world. Given B, it must remain a mystery as to why physicists are better at assigning prior probabilities than other groups — lawyers, theologians, etc. What is at issue here is not, of course, the failure of B to *justify* the choices of physicists in assigning prior probabilities to theories, but the failure of B to *make sense* of these choices, to render them intelligible or understandable. This can only be done by appealing to some (conjectured) feature or fact about the world which physicists, as a result of doing physics, have come (explicitly or implicitly) to believe in, but which lawyers or theologians may well not believe — such as that physicalism (in an appropriately strong sense) is true. Thus AOE can make sense of the evidence-independent choices of physicists, whereas B cannot: this, to my mind, is a devastating objection to B.

(2) Because B fails to make explicit metaphysical theses implicit in physicists' evidence-independent choices of theory, B cannot subject these influential metaphysical theses to rational, that is critical, scrutiny, in the light of potential and actual empirical fruitfulness, and other considerations, as AOE can. Whereas B must rely on the personal hunches and intuitions of physicists, AOE makes explicit, and so discussable, considerations relevant to the assessment of theories prior to evidence.

(3) The difference between AOE and SE just indicated makes it very unlikely that the (relatively) evidence-independent assessments of physicists proceeding in accordance with B will be as good as such assessments would be made by physicists proceeding in accordance with AOE. For AOE makes explicit how metaphysical theses (at levels 3 and 4) should *evolve* in the light of evolving knowledge at levels 1 and 2; granted B, assessments of prior probabilities can

amount to no more than the hunches of individual physicists.

(4) In so far as B has anything to say about factors that may influence physicists in assigning prior probabilities to theories, it is as inadequate as what other versions of SE have to say. Howson and Urbach (1993, pp. 417-418) refer to simplicity, and in particular to Jeffreys' account (Jeffreys and Wrinch, 1921), according to which the simplicity of a theory depends on the number of independent constants: but they acknowledge that this way of assessing simplicity is ambiguous.

(5) Like other versions of dressed SE (and unlike AOE), B can provide no rational method of discovery for fundamental theoretical physics.

(6) Howson claims that B is an extension of logic (Howson, 2000, chapter 7). This might lead one to believe that the Bayesian account of scientific method is more rigorous than that provided by AOE. Just the reverse is the case. AOE makes explicit, and B fails to make explicit, substantial, influential, problematic and implicit assumptions of science; AOE thus complies with, and B violates, the principle of intellectual integrity. All the other defects of B flow from this basic defect of lack of rigour.

B might, of course, have AOE adjoined to it; but then B would be redundant. AOE constrains acceptance of theories in a highly restrictive fashion, and nothing is achieved by scattering probabilities among its various theses. In short, B is either seriously defective, or redundant.

4 Rational Discovery of New, Fundamental Physical Theories

At the beginning of chapter two I gave a brief account of the rational, if fallible and non-mechanical, method of discovery which becomes available once SE is rejected and AOE is accepted instead. I went on to indicate that theoretical physics in practice is, in many ways, closer to AOE than SE when it comes to the development of new fundamental physical theories. This is strikingly apparent in Einstein's work in developing special and general relativity, and in the extraordinary contribution of Yang and Mills (1954), which led to the subsequent development of chromodynamics and the electroweak theory of Weinberg and Salam. And it is apparent in the massive amount of theoretical work devoted to the development of string (or M) theory, none of which has issued in a single successful empirical prediction. But if AOE is already implemented in scientific practice in this way, how could the explicit acceptance of AOE by the scientific community facilitate the discovery of new fundamental physical theories?

AOE would facilitate this by making explicit, and so criticizable and developable, implicit metaphysical assumptions.

A glance at the history of physics reveals that, again and again, physicists have taken for granted metaphysical ideas associated with precursor theories that are starkly at odds with the latest, best available theory. This has served to retard, not just the discovery of the latest theory, but its comprehension and acceptance once it has been discovered — to say nothing of the next revolutionary theory waiting to be discovered in the wings.

Thus, after the publication of Newton's *Principia* in 1687, Newton's theory was dismissed by Huygens and Leibniz for the simple reason that the law of gravitation is an action-at-a-distance theory, incompatible with the then current metaphysics of the corpuscular hypothesis. Huygens in a letter to Leibniz declares "Concerning the Cause of the flux given by M. Newton, I am by no means satisfied [by it], nor by all the other theories he builds upon his Principle of Attraction, which seems to me absurd . . . I have often wondered how he could have given himself all the trouble of making such a number of investigations and difficult calculations that have no other foundation than this very principle" (Koyré, 1965, pp. 117-118). In a sense, Newton agreed as he indicates in the remark: "That gravity should be innate, inherent, and essential to matter, so that one body may act upon another, at a distance through a vacuum, without the mediation of anything else . . . is to me so great an absurdity, that I believe no man who has in philosophical matters a competent faculty of thinking can ever fall into it" (Burtt, 1932, pp. 265-266). Attempts were even made to explain Newtonian gravitation in terms of the corpuscular hypothesis.

But gradually Newton's law won general acceptance. Eventually it led to the downfall of the corpuscular hypothesis and the development of a metaphysical blueprint in line with the spirit of Newton's law. Thus Richard Cotes defended the idea that gravitation is an innate property of all bodies (just the view that Newton had condemned as absurd) in his 1713 preface to the second edition of Newton's *Principia*. And later still, in 1763, Boscovich, in his *Theoria Philosophiae Naturalis* (one of the great works of AOE natural philosophy, published in English translation in 1922), expounded a new metaphysical blueprint, congenial to the spirit of Newtonian theory if not strictly compatible with it, according to which the world is made up of point-atoms each of which has mass and is surrounded by a rigid, spherically symmetrical field of force that varies with distance and is alternately repulsive and attractive as one moves away from the central point-particle. Not until over 60 years after the publication of Newton's *Principia* did a new blueprint emerge that accorded with the spirit of Newtonian theory.

In short, far from the current metaphysical blueprint aiding the discovery of Newtonian theory, it did exactly the opposite: it provided no rationale, no heuristics, for the development of the theory whatsoever and, furthermore, obstructed the interpretation, understanding, and even acceptance, of the theory. Over 60 years had to pass before a metaphysical blueprint was developed congenial to the new theory.

This pattern is repeated when it comes to the discovery, interpretation, understanding and acceptance of classical electrodynamics.

Something like Boscovich's blueprint came to prevail during much of the nineteenth century, so much so that some physicists defined the task of physics to be to explain phenomena along Boscovichean lines. Boscovich's blueprint embodied current ideals of physical intelligibility. Being widely accepted, it did much to block the development, acceptance and understanding of classical electrodynamics. The idea of the electromagnetic field as a new kind of physical entity, distinct from the corpuscle or Boscovichean point-particle, was invented by Faraday. The crucial difference between the electromagnetic field and the field of force surrounding the point-atom is that the latter is infinitely rigid, changes in the field travelling at infinite velocity, whereas the former is not rigid, changes travelling at the speed of light. Faraday was well aware of this crucial difference. He had little success, however, in communicating his ideas to his contemporaries. Instead of being regarded as one of the great theoretical physicists of all time for inventing one of the key concepts, Faraday was regarded — and is often still regarded — only as an experimental scientist of genius. It was James Clerk Maxwell, of course, who discovered the equations governing the classical electromagnetic field. But Maxwell used an elaborate mechanical model in deriving his field equations, and seems to have held that the field must be interpreted in terms of a more fundamental quasi-material aether, in turn to be understood as being made up, presumably, of something like Boscovichean atoms. The field idea was too violently at odds with the prevailing blueprint of the Boscovichean atom with its Newtonian-type rigid field of force, to be taken seriously as a new kind of physical entity, intelligible in its own right. After Maxwell, electrodynamics continued to be interpreted in this way, in terms of a hypothetical quasi-material aether. FitzGerald, Kelvin, Poynting, Heaviside, Lodge, Larmor and Lorentz all contributed to the development of Maxwell's theory interpreted in terms of the aether.[27] It was only with the advent of Einstein's special theory of relativity in 1905 that the aether dropped out of the picture, and the electromagnetic field began to be accepted as a new kind of physical entity, intelligible in its own right. But this Faraday-Einstein view was fiercely resisted; many regarded the idea of a field without the support of an underlying quasi-material aether as unintelligible.

We have here a pattern of confusion remarkably similar to that indicated in connection with Newtonian theory. Far from the current (Boscovichean) blueprint aiding the discovery and development of Maxwellian theory, it did exactly the opposite: it provided no rationale, no heuristics, for the development of the theory whatsoever and, furthermore, obstructed the interpretation, understanding, and even acceptance, of the theory. Nearly half a century passed before a view emerged congenial to the new theory, interpreting the theory as being about a new kind of physical entity.

A similar, if rather more controversial, story can be told in connection with the development of quantum theory. It is reasonable to conjecture that the transition from classical to quantum physics involves a transition from determinism to probabilism. If this conjecture is correct, then quantum theory requires a new probabilistic blueprint. No such blueprint emerged during the period in which quantum theory was created, roughly from 1900 to 1926. After the discovery of the wave aspects of the electron, physicists despaired of understanding quantum entities as either particles, or waves (i.e. in terms of a field). But if nature is fundamentally probabilistic, this was exactly the wrong thing to try to do. For the classical particle and wave are deterministic entities,

---

27. See (Hunt, 1991) and, at a somewhat more technical level, (Buchwald, 1985), for excellent accounts of the development of classical electrodynamics after Maxwell.

and these are entirely inappropriate if the quantum domain is fundamentally probabilistic. What is required is the invention of a new kind of unproblematic, fundamentally probabilistic entity, a new kind of probabilistic blueprint. For a suggestion along these lines, leading to a new version of quantum theory, free of the theoretical problems that plague orthodox quantum theory, see (Maxwell, 1972; 1976b; 1982; 1988; 1994b; 1998, ch. 7; 2004c).

Here again, metaphysical ideas (determinism, the classical deterministic particle and field), far from encouraging the development of the new theory (quantum theory), do exactly the opposite: they provide no rationale, no heuristics, for the development of the new theory whatsoever and, furthermore, obstruct the interpretation, understanding, and even acceptance, of the theory. As I have argued at length, orthodox quantum theory is *ad hoc*, imprecise, ambiguous, non-explanatory, restricted in scope and resistant to unification because of its failure to solve the wave/particle problem and the resulting need to interpret the theory as being about the results of performing measurements on quantum systems (see above references). Over 70 years after the creation of quantum theory, these issues have still not been resolved.

What these three major episodes in the history of theoretical physics reveal is that accepted metaphysical ideas, instead of aiding the discovery of new fundamental theories, block the discovery, acceptance, interpretation and understanding of new theories as a result of being dogmatically upheld even though appropriate only to predecessor theories or theoretical circumstances. Theoretical physicists appear to be brilliantly innovative when it comes to equations, but conservative, unimaginative and dogmatic when it comes to metaphysics.

But this is exactly what one would expect granted general acceptance of SE, or at least general failure to put AOE into practice!

The crucial question is: Do we have any reason to suppose physicists would have done better if AOE had been generally accepted, from Galileo's time onwards, let us suppose?

I claim we do. The general point is that AOE demands that implicit metaphysical assumptions are made explicit within a framework designed to promote sustained imaginative and critical scrutiny of such assumptions, and those rival assumptions most likely to be fruitful. If versions of SE are accepted instead of AOE, versions which banish metaphysics from science, then metaphysical assumptions will be made surreptitiously, unconsciously, and as a result are likely to be held dogmatically, and to be outdated. Furthermore, repeated failure of outdated metaphysics to provide intelligible interpretations of new theories will be held to be strong grounds for regarding metaphysics as at best irrelevant, at worst harmful, to physics. Anti-realist views (whether   taking   the   form   of   instrumentalism,   positivism, conventionalism or constructive empiricism) will become fashionable, further reinforcing SE and undermining any movement towards AOE. But this of course misses the point entirely. For it is outdated, dogmatically upheld metaphysics that is harmful to physics, not metaphysics as such. The lesson to be learnt is the exact opposite of anti-realism. What is required is sustained imaginative and critical exploration of metaphysical possibilities carried on as an integral part of physics itself.

A much more specific and powerful point is this. Elsewhere I have shown how, beginning with the corpuscular hypothesis, ignoring empirical considerations, and taking at most physicalism for granted as a background assumption, one can generate the whole sequence of blueprints, from the corpuscular hypothesis to Boscovich's point-atomism, to a field/particle blueprint, to an Einsteinian self-interacting unified field blueprint incorporating basic ideas of special relativity, merely by modifying pre-existing ideas in order to solve problems inherent in these ideas (see Maxwell, 1998, ch. 3). I have also argued that deterministic blueprint ideas, subjected to the demand to be generalized to accommodate probabilism can, similarly, generate blueprint ideas appropriate to quantum theory, and capable of leading to a new version of quantum theory experimentally distinct from the orthodox version of the theory, but as yet untested (Maxwell, 1988, 1994b, 1998 ch. 7, 2004c).

This provides a powerful case for holding that theoretical physics would have been even more successful if AOE had been put into practice from Galileo's time onwards.

The great exception to the pattern indicated above, of outdated metaphysics obstructing acceptance of a new theory rather than new metaphysics, rationally developed, aiding the discovery of the new theory, is Einstein's discovery of general relativity (referred to, briefly, in chapter two). Einstein first discovered the blueprint — the idea that gravitation is a feature of curved spacetime; only some years later did he discover how to capture this idea in the equations of general relativity.

But what about today? Given that string theory is pursued without issuing in any successful empirical predictions, does not this indicate that AOE is now being put fully into practice? It does not, as we saw briefly in chapter two. String theory was discovered and developed as a result of attempts to develop methods for deriving predictions from theories. It began life when a theoretical physicist, Gabriele Veneziano, discovered in 1968 a mathematical technique for predicting the results of collisions between hadrons — particles like protons and neutrons (which go to make up the nuclei of atoms). Two years later it was realized, by Yoichiro Nambu and others, that the new mathematical technique could be interpreted as asserting that protons and neutrons (and other hadrons), instead of being point-particles, are tiny strings. But, after its initial success, this new theory of strings was found to have serious empirical problems. And in the meantime, a much more successful rival theory had been developed. This built on an idea proposed in 1964 by Murray Gell-Mann, and independently by George Zweig, according to which protons and neutrons are made up of new kinds of particle called quarks. By the mid-1970s a quantum theory of the force between quarks had been developed, namely chromodynamics. This theory met with empirical success, and string theory was forgotten, except for two physicists, John Schwartz and Joel Scherk, who reinterpreted string theory as a theory about all fundamental particles, a theory which, to them, seemed to promise to unify quantum theory and general relativity. Eventually, in 1984, Schwartz and Michael Green showed that this new string theory of quantum gravity could be used to make sensible calculations, when all rival approaches to quantum gravity yielded absurd infinite results. String theory abruptly hit the headlines. Theoretical physicists started to work furiously to develop the new string theory.

Even this highly condensed early history of string theory reveals that the way the theory was discovered and developed differs strikingly from the method of discovery of aim-oriented empiricism — which involves attempting to get at the heart of the clash between accepted level 2 theories and accepted level 3 and level 4 assumptions (of Fig. 1.2), by extracting basic, clashing ideas or principles from each theory, and then modifying level 3 ideas to achieve greater unity, these ideas then being made more precise until they become a new, testable physical theory.

There is still a failure to explore blueprint possibilities shorn of detailed mathematical technicalities. String theory has not come to grips with the profound theoretical inadequacies of quantum mechanics. Probabilistic metaphysics remain unarticulated and unexplored. There exists no range of clearly articulated metaphysical options that constitute different possible solutions to the problem of unifying GR and quantum theory, GR and SM. Even the task of articulating coherent possible probabilistic dynamic geometries does not seem to have been undertaken. Thinking like Einstein's, or of the kind that Faraday engaged in, bereft of mathematical technicalities, leading to the development of the fundamental idea of the classical field, appreciated by James Clerk Maxwell but by few others, does not take place explicitly in journals, no doubt because it would be judged to be too naïve, too primitive, to be of any value. Alternatives to physicalism are not articulated and assessed. In the next section I attempt to repair this last omission by considering one such alternative.

5 A Possible Alternative to Physicalism

Modern physics, and physicalism, can be understood as emerging from one kind of response to Parmenides. The rival to physicalism I wish to consider arises from a rather different response to Parmenides.

According to Parmenides, the universe is an unchanging, homogeneous sphere. All change and diversity is illusory. This can be seen as one of the first, and most extreme, versions of the idea that we live in a physically comprehensible universe. Parmenides seems to have held this extraordinary view because he thought that the very idea of change or motion involves a contradiction. An object can only move if there is an empty space, a region of nothingness, into which it can move. But this in turn requires that nothingness exists, which amounts to holding that the non-existent exists, a contradiction. Hence there can be no motion.

Democritus took the bull by the horns, and argued that there is motion, and hence the nothingness must exist after all. He imagined nothingness surrounding Parmenides' unchanging sphere. In this nothingness other Parmenidean spheres might exist which, when shrunk down to a minute size, became atoms. Thus was born one of the most fruitful scientific theories ever conceived. Richard Feynman once declared "the world is made of atoms" to be the single most important sentence of modern science.

Atomism gave rise to Boscovichean point-atomism, to Faraday's and Einstein's idea of the field, to later blueprint ideas, and to the idea that what is unchanging and invariant is not any actual state of the universe, but rather that which corresponds physically to laws governing phenomena, necessitating physical properties as I have argued elsewhere (Maxwell, 1968; 1998, pp. 141-155).

But another reply to Parmenides is possible. We may declare that Parmenides' homogeneous sphere is a state of the entire universe, exhibiting unity, at a very special time, namely the moment before the big bang. Before the big bang, the entire universe is in a state of extreme unity, all disunity being merely virtual, somewhat like the virtual particle creation and annihilation processes which go on in the vacuum according to quantum field theory (on one popular reading of the theory at least). Then the big bang occurs, an instant of spontaneous symmetry breaking: the outcome is a multitude of virtual prior-to-big-bang states, virtual Parmenidean spheres, as it were. The subsequent history of the cosmos is the unfolding of the interactions between these multitudinous virtual prior-to-the-big-bang entities. All the change and diversity that exists in the world around us is the outcome of the diverse ways in which virtual Parmenidean "spheres" are inter-related with one another.

We have here, then, two versions of physicalism. There is the version that has emerged from Democritus' response to Parmenides, hitherto called physicalism, and which in what follows I will call "atomistic physicalism". And there is the version which emerges from the response to Parmenides just indicated, which I shall call "cosmic physicalism".

Atomistic and cosmic physicalism give diametrically opposed answers to the question: What is the nature of the physically simple or elemental? The first declares this to be either empty space, or the interior of the atom (as far as Democritus' version of physicalism is concerned); the second declares this to be a special state of *everything*!

These two versions of physicalism can be distinguished a bit more formally as follows. Granted that physicalism divides physical reality into two parts, U, that which does not vary, and V, that which does vary, we may stipulate that all versions of physicalism specify a special "Parmenidean" state of the universe, such that V does not vary either, a state which is such that the symmetries of U are also symmetries of V.

(Or it might be that **U** and **V** taken together exhibit symmetries with a group structure distinctive of unity, that is the symmetry group is not a product of subgroups. Or **V** might disappear altogether, leaving only **U**.)

There are now two possibilities:

(I) *Atomistic physicalism.*   The Parmenidean state is a conceptual possibility only: it only occurs in actuality in restricted spatial regions (where there are no atoms, or inside any Democritean atom), or perhaps only approximately (in empty space where forces are almost zero).

(II) *Cosmic physicalism.*   The Parmenidean state is an actual special state of the entire universe.

The corpuscular, point-atom and field blueprints are all versions of *atomistic* physicalism.   Atomistic physicalism has been the dominant basic idea in the history of theoretical physics so far.   It is possible, however, that it is *cosmic*, rather than atomistic, physicalism which is true.   The following suggestive developments in theoretical physics during the 20th century may well be interpreted as pointing towards cosmic physicalism.

(1) A basic idea of atomistic physicalism is that the physically simplest, most elemental state of affairs that can exist is the vacuum: physical states of affairs become progressively more complex as 1, or 2, or ... n atoms, or fundamental particles, are added to the vacuum.   With the advent of field theory, however, this straightforward ordering of physical complexity begins to break down: here, presumably, we would have to say that the simplest state obtains when the value of the field is everywhere zero, a less simple state arising when the value of the field is everywhere a constant value, more complicated states arising with increasingly complicated variable values of the field.

(2) The idea that the vacuum is the simplest state, in that it is always present in an unchanging form, begins to break down with the advent of Einstein's general theory of relativity (GR).   According to GR, the curvature of space-time varies with varying amounts of matter, mass or energy-density; and curved spacetime itself possesses energy. Space is no longer a bland, unchanging arena within which more or

less complex physical events unfold: the variable curvature of space, or of spacetime, itself takes part in dynamical evolution. (Empty, flat, unchanging space-time still arises, however, as a possible solution to the equations of GR.)

(3) With the advent of quantum field theory (QFT), empty space becomes even more complex in that it is full of so-called vacuum fluctuations. These may be pictured as follows. According to one (perhaps somewhat dubious) way of interpreting the time-energy uncertainty relations, $\Delta t \Delta E \geq h/2\pi$ (here, h is Planck's constant), given any state of affairs, for very short time-intervals, Δt, there will be an uncertainty of energy, ΔE, with $\Delta E \approx \Delta t.h/2\pi$. Thus, even in empty space, there is sufficient energy, ΔE, available to create an electron/positron pair anywhere, at any time, with the proviso only that such a pair must mutually annihilate after a time Δt, with $\Delta t < \Delta E.h/2\pi$ and $\Delta E > 2m_e c^2$, where m_e is the rest mass of the electron. Such fleetingly existing particles are called "virtual" particles. According to QFT, each minute space-time region is full of virtual processes, involving the creation and annihilation of particle/anti-particle pairs. Indeed, all possible virtual processes occur that violate no other conservation law except that of energy (understood classically). Within very tiny space-time regions so much (virtual) energy can exist that there is sufficient to create a short-lived virtual black hole (a point that will be picked up again below).

According to QFT, in other words, the vacuum is a mass of seething activity, which averages out to nothing over sufficiently large space-time regions. One may, indeed, interpret QFT as a theory of the vacuum, all possible physical processes going on, within minute space-time regions, as *virtual* processes. In supplying discrete units of energy, ΔE_1, ΔE_2, ... ΔE_n, we merely change some of the *virtual* processes into *actual* processes. There is a sense in which the most complex physical state imaginable is no more complex than the vacuum: it is just an energetic state of the vacuum such that some virtual processes are actual processes.[28]

In brief, QFT transforms the simple, elemental, unchanging vacuum of 19th century physics into a seething mass of complex, virtual processes — mirroring, in a ghost-like way, all the complexity of the most complex actual physical processes that exist when there

---

28. For an excellent account of the quantum theory of the vacuum, see (Aitchison, 1985).

is matter. It might be supposed that quantum vacuum fluctuations are not real physical phenomena, but only artefacts of the formalism of QFT given a certain (questionable) interpretation. But this does not take into account that vacuum fluctuations have been detected! This was done decades ago by means of the Casimir effect. According to QFT, if two metal plates are held a small distance apart, virtual processes that involve the creation and annihilation of electron/positron pairs will tend to be *suppressed* in the space between the plates. This results in there being a small pressure tending to push the plates together, due to the *unsuppressed* virtual processes taking place in the space surrounding the two plates. This minute force, due to vacuum fluctuations — the Casimir effect — has been detected and measured.

(4) Cosmic physicalism requires that there is a special state of the entire cosmos which is such that all diversity and change disappears: the very distinction between space and matter, we may presume, disappears, there being just *one* homogeneous, instantaneously unchanging *something*, which is also *everything*. This is made possible by big bang cosmology. It is conceivable that the big bang state of the universe, when all of space and matter was packed into a tiny region, was the Parmenidean state of instantaneous unity and homogeneity.

(5) According to GR, the force of gravity is not a force at all; it is rather the tendency of the curvature of space-time to be affected by the presence of matter, or energy-density. With GR, one apparent force, gravity, becomes a feature of physical geometry. This suggests that it may be possible to carry this process of "geometricizing" physics further, the eventual outcome being the unification of space-time, on the one hand, and matter or energy on the other hand. Just this is required by cosmic physicalism.

(6) According to the Salam-Weinberg theory of the electroweak force (QEWD), at high energies, the distinction between the electromagnetic force on the one hand, and the weak force on the other, disappears. As we go backwards in time towards the big bang, towards a time when the energy-density of the universe was sufficiently high, there existed just one unified force, the electroweak force. As the universe expanded, and the energy-density went down, the unity of the electroweak force was broken:

the currently observed disunity of two forces with very different properties emerged. This is strikingly in accord with the basic idea of cosmic physicalism: as we move backwards to the original big bang state so we move towards a state of affairs of greater unity, simplicity, symmetry or homogeneity. The potentially immensely important idea of cosmic spontaneous symmetry breaking, which the Salam-Weinberg theory introduces at the level of fundamental theoretical physics, is precisely what is required to make cosmic physicalism a possibility. It is this development in theoretical physics, above all, which makes cosmic physicalism a viable possibility.

(7) The idea of an initial state of high symmetry or unity evolving into something asymmetrical and disunified, is further supported by superstring theory. According to this theory, space has 10 or 26 dimensions (or 11 dimensions according to M theory). The dimensions that we do not observe are curled up into such a minute multi-dimensional "ball" that we do not ordinarily notice their existence. The idea here is that at the big bang state all the 10, 11 or 26, dimensions of space were curled up in this fashion; the subsequent evolution of the universe consists of just three spatial dimensions growing in size to become, eventually, the space in which we find ourselves. Here, an original Parmenidean unity becomes disunity as three dimensions of space become dramatically different from the rest.

Here, then, are seven developments in theoretical physics and cosmology which took place during the 20th century which may be taken to be steps towards cosmic physicalism, indications that it is in this direction that theoretical physics is progressively developing.

What would a theory of everything, T_C, that accords with cosmic physicalism look like? How would it differ from a theory of everything, T_A, that accords with atomic physicalism?

The major difference between any T_A and any T_C would be that they specify diametrically opposed conditions for the simple, elemental, unified or homogeneous to exist. For T_A this is the vacuum, the state with as little as possible; for T_C this is a special state of *everything*, of the entire *cosmos*.

Another important difference is that, for any T_A, there is a fundamental difference between theory and initial conditions, even when the theory is applied to the cosmos as a whole. For all such applications, initial conditions must be specified in addition to the theory, and are not supplied by the theory itself. According to any T_C, however, the theory itself specifies a cosmic state of extreme unity. The initial state is, in a sense, given by the theory itself. It might even be the case that the theory is *only* applicable if the unique cosmic state of unity exists; universes for which this is not the case are, according to the theory, not possible. In this case the theory would specify, and assert the existence of, the initial cosmic state in a very strong sense.

Another important difference stems from the fact that, whereas T_A is not obliged to postulate an actual cosmic state of unique unity, T_C *is* obliged to postulate such a state. If we take this state to be in the past, then T_C is, in part, a theory about how disunity *develops* in time. T_C must imply that the cosmos evolves from an initial state of unity into states of increasing disunity, as a result of episodes of *spontaneous symmetry breaking*. First principles do not require of any T_A that it provide such a cosmic history of increasing disunity. T_C, unlike T_A, is an inherently cosmological and historical theory.

Any T_C will only be able to predict a cosmic evolution of increasing disunity if it is fundamentally *probabilistic* in character. A deterministic T_C could only specify how an initial minute, or implicit, disunity becomes large or explicit with the passage of time, since the eventual cosmic diversity that we experience today would be precisely determined from the outset. Spontaneous symmetry-breaking, in other words, is an inherently *probabilistic* event (one asymmetrical state of affairs being selected probabilistically from a number of other asymmetrical possibilities). Any T_C must, then, be a fundamentally *probabilistic* theory, something that is not required, *a priori* as it were, from any T_A. (The only way to avoid this consequence is to hold that the universe today is entirely unified, disunity or diversity only being apparent: this is, of course, Parmenides' own view. In an apparently disunified and probabilistic world it gets rid of probabilism in somewhat the way in which the many-worlds, or Everett interpretation of quantum theory (QT) gets rid of probabilism.)

In the light of this argument, we may hold that the fundamentally probabilistic character of QT provides an eighth suggestive pointer towards cosmic physicalism. (This requires of course that we interpret

QT as being fundamentally probabilistic, in the first place. It also requires that QT can be interpreted in such a way as to associate fundamentally probabilistic events with spontaneous, cosmic symmetry-breaking.)

The considerations developed so far strongly suggest that the universe might be physically comprehensible in the following manner. If we take seriously that, within any small spatial region, energy ΔE is available for time intervals Δt, where $\Delta E \approx h/2\pi\,\Delta t$, we will be led to conclude that *at any space-time point*, (x,t), infinite energy will exist. Granted that some T_C is true, however, at any space-time point, (x,t), or rather any sufficiently tiny space-time region around what we may construe to be a space-time point (x,t), only the energy equivalent of the entire cosmos can exist. In other words, we may hold that, at each space-time point, the original big bang state of unique unity *exists as a virtual state*. Initially, only the big bang state of complete unity exists. This is an *actual* state. Then the *actual* state of the cosmos becomes a superposition of *instantaneous and point-like virtual cosmic states of complete unity*. It is the existence of the original cosmic state of complete unity as a *virtual* state at each space-time point that ensures that the variable *actual* state of the cosmos evolves as it does. In short, according to this quantum version of cosmic physicalism, the **U** that exists everywhere, determining how the variable **V** evolves, is the Parmenidean state of cosmic unity, as a *virtual* state at each space-time point. It is the existence of the unified cosmic state at each space-time point, as a virtual state, that ensures that the variable changes as it does.

Initially, there is only unity, and all disunity is purely virtual. Then unity becomes virtual, and disunity actual, as a result of cosmic, probabilistic symmetry-breaking.

6 Does Aim-Oriented Empiricism Solve the Problem of Induction?

6.1 Importance of the methodological part of the problem of induction

There are three parts to the problem of induction:

(1) *The methodological problem of induction.* What methods enable us to select physical laws and theories on the basis of empirical considerations in a way which does justice to what goes on in

science?

(2) *The theoretical problem of induction.* What justification can there be for accepting any physical law or theory, however empirically successful, granted that our aim is to improve our theoretical knowledge and understanding of the universe?

(3) *The practical problem of induction.* What justification can there be for accepting any physical law or theory, however empirically successful, granted that our aim is to use the law or theory as a basis for action?

As far as (1) is concerned, two kinds of method are employed to select theories in physics: methods concerning *empirical* requirements accepted theories must satisfy, and methods concerning *non-empirical* requirements, having to do with such things as *unity* and *simplicity*. Specifying *empirical* methods does not pose too much of a problem; it is specifying *non-empirical* methods that has seemed to be profoundly problematic, because of the mystery as to what the *unity* or *simplicity* of a theory is.[29] But, as we saw in section 2 above, AOE solves this problem. In doing so, AOE solves problem (1).

This may seem to be a very meagre achievement. All that is involved, in solving (1), is the specification of the *methods* that are employed in physics in the acceptance and rejection of theories. There is no claim that theories, accepted by means of these methods, constitute contributions to knowledge, in some worthwhile conception of knowledge.

The achievement may seem less meagre, however, when one takes the following into account. Solving (1) is an essential preliminary to solving (2) and (3). If we cannot even specify the methods employed in accepting and rejecting physical theories, it is unlikely that we will be able to justify the claim that these (unspecified) methods deliver knowledge, in some worthwhile sense of knowledge.

Ever since Hume highlighted the problem, and Kant recognized just how serious the problem is, in 1738 and 1781,[30] philosophers have struggled to solve it. A vast literature exists on the subject: see, for

---

29. SE may create the false impression that empirical methods are problematic because it gives far too big a role to empirical considerations in science, and far too small a role to non-empirical considerations, because of the problematic character of the latter.

30. I allude, here, to (Hume, 1959) and (Kant, 1961).

example (Kyburg, 1970; Swain, 1970; Watkins, 1984; Howson, 2002 — and references given therein). Nowadays, philosophers despair of solving the problem, and it has almost fallen into disrepute. But there is a simple explanation as to why all these attempts to solve the problem have failed: they have in effect tried to justify the methods of science — as methods that deliver trustworthy knowledge — when in fact the methods presupposed have been *invalid*, being those of some version of SE (or at least not those of AOE). The vast literature attempting to solve the problem of induction has failed, in other words, because it has sought to justify the unjustifiable, show to be rational what is irrational. This is of course symptomatic of the neurosis of science, as I pointed out in chapter one.

Thus we have ample grounds here for holding that solving problem (1) is far from a meagre matter. It is just the traditional concentration on (3), to the neglect of (1) — resulting in failure to solve (1) — which has been responsible for the failure of all these efforts. I would go further: (1) is really the key to solving (2) and (3). Once (1) has been solved, solving (2) and (3) involves no more than adjusting our ideas in the light of what we have learned from the solution to (1).

6.2 Solution to the circularity objection

The claim that AOE solves the problem of induction is immediately confronted with the objection that AOE has vicious circularity built into it. Acceptance of physical theories is influenced by their degree of accord with metaphysical principles, the acceptance of which is in turn, in part, influenced by an appeal to the empirical success of physical theories. The claim is that as theoretical knowledge and understanding improves, metaphysical theses and associated methods improve as well. There is something like positive feedback between improving knowledge, and improving knowledge–about–how–to–improve knowledge. That, I claimed in section 1.3 of chapter three, is the methodological key to the great success of modern science, namely that it adapts its metaphysical assumptions and methods (its aims and methods) to what it finds out about the nature of the universe. But how can such a circular procedure conceivably be valid?

A number of points need to be made in response to this objection.

First, accepted level 2 theories are actually *incompatible* with metaphysical principles at levels 3 or 4. There is no question of justifying the truth of physical theories by an appeal to metaphysics, the truth of which is in turn justified by an appeal to the truth of physical theory.

Second, I endorse Popper's thesis that all our knowledge is ultimately conjectural in character. We may justify *acceptance* of a theory or metaphysical thesis, or *preference* for one theory or metaphysical thesis over another, but no attempt is made to justify a claim such as that we know for certain, or with such and such a degree of probability, that such and such a theory or metaphysical thesis is *true*.

Third, it deserves to be appreciated that the two-way influence, between level 2 theories and metaphysical theses at levels 3 to 5 at least, is mirrored by a similar two-way influence that occurs when theory and experiment clash. In general, if a theory clashes with an experiment that has been subjected to expert critical scrutiny and repeated, the theory is rejected. But on occasions it turns out that it is the experimental result that is wrong, not the theory. In a somewhat similar way, if a new theory increases the conflict between the totality of physical theory and the currently accepted metaphysical thesis, at level 3 of Fig. 1.2 of chapter one, the new theory will be rejected (or not even considered or formulated). On occasions, however, a new theory may be developed which increases the conflict between the totality of theory and the current thesis at level 3 but decreases the conflict between the totality of theory and physicalism at level 4 of Fig. 1.2. In this case the new theory may legitimately be accepted and the thesis at level 3 may be revised. In principle, as I have already indicated, theses even higher up in the hierarchy may legitimately be revised in this way. A virtue of the hierarchical view is that it makes possible and facilitates such two-way revision.

Fourth, the above circularity objection is not valid when made against the modest version of AOE, expounded in chapters one and two, which claims to be more rigorous than standard empiricism because implicit metaphysical theses are made explicit, but which does not claim to justify claims to knowledge and solve the problem of induction.

But fifth, the moment we go beyond this modest position, and claim that AOE does solve the problem of induction, accepted theories being the best available attempts at capturing explanatory truth, the circularity objection does inevitably arise. How can it possibly be valid to justify acceptance of theories by an appeal to metaphysical principles, and then justify acceptance of these principles by an appeal to the empirical success of theories?

Here, in a nutshell, is the answer. Permitting metaphysical assumptions to influence what theories are accepted, and at the same time permitting theories to influence what metaphysical assumptions are accepted, may (if carried out properly), *in certain sorts of universe*, lead to genuine progress in knowledge. The level 6 thesis of meta-knowability of AOE asserts that *this is just such a universe*. And furthermore, crucially, reasons for accepting meta-knowability make no appeal to the success of science. In this way, meta-knowability legitimises the potentially invalid circularity of AOE.

Relative to an existing body of knowledge and methods for the acquisition of new knowledge, possible universes can be divided up, roughly, into three categories: (i) those which are such that the meta-methodology of AOE can meet with no success, not even apparent success, in the sense that new metaphysical ideas and associated methods for the improvement of knowledge cannot be put into practice so that success (or at least apparent success) is achieved; (ii) those which are such that AOE appears to be successful for a time, but this success is illusory, this being impossible to discover during the period of illusory success; and (iii) those which are such that AOE can meet with genuine success. Meta-knowability asserts that our universe is a type (iii) universe; it rules out universes of type (i) and (ii).

Meta-knowability asserts, in short, that the universe is such that AOE can meet with success and will not lead us astray in a way in which we cannot hope to discover by normal methods of scientific inquiry (as would be the case in a type (ii) universe). If we have good grounds for accepting meta-knowability as a part of scientific knowledge – grounds which do not appeal to the success of science – then we have good grounds for adopting and implementing AOE (from levels 5 to 2).

But what grounds are there for accepting the thesis of meta-knowability at level 6? These were spelled out above in section 3 (p. 176). They are not, perhaps, very strong grounds for accepting meta-knowability, and are open to criticism. But the crucial point, for the present argument, is that these grounds for accepting meta-knowability are independent of the success of science. This suffices to avoid circularity.

If AOE lacks meta-knowability, its circular procedure, interpreted as one designed to procure justified knowledge, becomes dramatically invalid, as the following consideration reveals. Corresponding to the succession of accepted fundamental physical theories developed from Newton down to today, there is a succession of *ad hoc* rivals which postulate that gravitation becomes a repulsive force from the beginning of 2050, let us say. Corresponding to these *ad hoc* theories there is a hierarchy of *ad hoc* versions of physicalism, all of which assert that there is an abrupt change in the laws of nature at 2050. The *ad hoc* theories, just as empirically successful as the theories we accept, render the *ad hoc* versions of physicalism just as scientifically fruitful as non-*ad hoc* versions of physicalism are rendered by the non-*ad hoc* theories we actually accept. If we take it as given that we accept non-*ad hoc* theories, the question of what reasons there are for rejecting empirically successful *ad hoc* theories and associated *ad hoc* versions of physicalism does not arise. But the moment we seek to *justify acceptance* of non-*ad hoc* theories and *rejection* of *ad hoc* theories, within the framework of AOE, the question of what reasons there are for rejecting *ad hoc* theories and associated *ad hoc* versions of physicalism arises. If AOE is bereft of meta-knowability, it is not easy to see what these reasons can be. But AOE with meta-knowability included does provide a reason: the *ad hoc* versions of physicalism assert that this is a type (iii) universe, which violates meta-knowability.

6.3 How is the problem of induction to be formulated?

In seeking to solve parts (2) and (3) of the problem of induction, it is important to avoid trying to solve versions of the problem that make epistemological demands that are so strong that they are quite impossible to fulfil, and the problem, so formulated, is insoluble. What we need to do is solve that version of the problem which makes the strongest achievable epistemological demands.

The problem might be formulated as: How can our confidence that empirically successful, accepted scientific theories are true, be justified? This makes impossibly strong epistemological demands. All physical theories so far put forward, whatever empirical success they may have achieved, are false![31] Formulated in this way, the problem is insoluble.

---

31. This point was made at the beginning of chapter two, and above in section 3. When

A slightly less epistemologically ambitious formulation would be: How can our confidence that the empirical predictions of empirically successful, accepted theories are true, be justified?" But this too asks for too much. All physical theories so far proposed, however empirically successful, yield false empirical predictions. A still less epistemologically ambitious formulation would be: How can our confidence that empirically successful, accepted theories yield true empirical predictions, within the standard range of phenomena (and accuracy) for which they have already been shown to be successful, be justified? But even this may ask for the impossible. Perhaps our customary confidence in science is misplaced. Perhaps just this is revealed by the correct solution to problems (2) and (3).

In short, in order to avoid struggling to achieve the impossible, we need to formulate the problem in a somewhat more open-ended way than any of the above. The following stands a better chance of being solvable: How can our confidence that empirically successful, accepted theories yield true empirical predictions, within the standard range of phenomena (and accuracy) for which they have already been shown to be successful, be justified *in so far as such confidence is justified*?

We cannot just assume, from the outset, that the solution to the problem must justify our "pre-Humean" confidence in common sense and scientific knowledge — the confidence we had, that is, before learning of Hume's devastating arguments. For this may, again, be asking for the impossible. Perhaps Hume demonstrated, decisively, that such "pre-Humean" confidence is misplaced and unjustifiable. The correct solution would, in this case, demonstrate just this to be the case. This, of course, is the view defended here.

6.4 The problem of induction as a part of general scepticism about knowledge

Parts (2) and (3) of the problem of induction arise essentially because

---

viewed from the standpoint of SE, the fact that science advances from one false theory to another, seems discouraging, and is often called "the pessimistic induction". Viewed from an AOE perspective, this mode of advance is wholly encouraging, since it is required by AOE. Granted physicalism, the only way a dynamical theory can be precisely true of any restricted range of phenomena is to be true of all phenomena. All physical theories must be false until we obtain a theory of everything!

of Hume's sceptical arguments which seem to show that knowledge of physical laws (and knowledge about the future) is impossible. When construed in this way, the problem of induction is part of a more general problem of how to combat sceptical arguments which seem to show that knowledge, of one kind or another, is impossible. There are, for example, sceptical arguments which seem to show that we cannot have knowledge of the material objects we experience, since our senses may systematically deceive us (we can never compare our experiences of objects with objects themselves, but can only compare one experience with another). Likewise, there are sceptical arguments which seem to show we cannot have knowledge of the past, since our memories and records of the past may systematically deceive us (we can never compare our memories and records of the past with the past itself, for the past is a country we cannot revisit). A part of what is involved, then, in solving the problem of induction, is producing a general response to various sceptical arguments about knowledge.

There is a long-established tradition in philosophy, going back at least to Descartes and Locke, which seeks to rebut scepticism by finding some especially secure part of what we ordinarily take to be knowledge which is immune to doubt. One then seeks to build up the rest of knowledge on this secure foundation. For Descartes, the very act of doubting provides this secure foundation: "I doubt, therefore I am". For Locke, and for empiricist successors such as Berkeley, Hume, Mill and Russell, the secure foundation is immediate experience (sense impressions or sense data).

Popper turned this tradition on its head. The fundamental problem of knowledge, for Popper, is the problem of how to *improve* what we take to be knowledge. In order to set about *improving* knowledge we need to give priority, not to what is immune to doubt, but just the opposite, to what is most vulnerable to doubt, to criticism. We learn, we improve our ideas, through discovering our mistakes, through detecting and eliminating error. Hence we need to give priority to those claims to knowledge which are most vulnerable to being shown to be false, namely empirically falsifiable theories. Science makes such staggering progress in improving knowledge, not because it gives priority to secure foundations immune from doubt, but just the opposite, because it gives priority to speculations open to the most devastating form of doubt, of criticism, possible, namely empirical refutation. And more generally, even when claims to knowledge are not empirically falsifiable, they can still be *criticized* if put forward as conjectural attempts at solving problems of knowledge. Scepticism, for Popper, is not the enemy of reason, to be defeated if reason is to prevail; quite the contrary, it is the heart of reason. We improve knowledge by subjecting what we take to be knowledge to persistent sceptical attack.

Traditional sceptical arguments acquire their force from putting forward irrefutable conjectures which if true, render great swathes of what we ordinarily take to be knowledge to be wholly and utterly false. Thus scepticism about perception puts forward the irrefutable conjecture that what we take to be knowledge of things around us based on perception is utterly false — perception deluding us in a persistent, consistent way. Scepticism about the past puts forward the irrefutable conjecture that all our memories and records of the past have been created to make it seem they give us self-consistent knowledge of the past whereas actually no such past exists at all. And Humean scepticism about theoretical knowledge, or knowledge of the future, puts forward the now irrefutable conjecture that the course of nature will change abruptly in the future, and all our current theories and expectations will turn out to be utterly false. The traditional approach to defeating scepticism, in order to establish a base of certainty (or high probability), has to find some way either of refuting these irrefutable conjectures, or of showing that these conjectures are of limited scope so that, even if true, the base of certainty is not undermined.

Viewed from a Popperian perspective, the problem of scepticism looks different. Irrefutability, far from being a virtue, is a vice. The irrefutable knowledge-destroying conjectures of scepticism do not deserve to be taken seriously, not because we can show them to be false after all, but precisely because they are irrefutable. In seeking to improve knowledge, we need to give priority to *refutable* conjectures; the irrefutable conjectures of scepticism deserve to be dismissed precisely because they are irrefutable.

But, as we have seen, this Popperian approach fails when it comes to Humean scepticism about theoretical knowledge. *Ad hoc* rivals to accepted theories may be even more falsifiable than accepted theories.

It fails because Popper did not go far enough. It is not enough to be critical. We need to be *critically* critical. That is, we need to be critical of criticism itself. The whole point of criticism, from a Popperian standpoint, is to make a contribution to the growth, the improvement, of knowledge. Scepticism, criticism, is *rational* only as long as it holds out some promise, some possibility, of contributing to the improvement of knowledge. Criticism which cannot contribute to the improvement of knowledge ceases to be rational, and deserves to be dismissed on that account. Furthermore, it is not enough to be *indiscriminately* critical; we need to direct criticism to those parts of what we take to be our knowledge which, we conjecture, are the most fruitful to criticize — which are such as to hold out the greatest hope of promoting the improvement of knowledge when subjected to criticism. Criticism which cannot contribute to the improvement of knowledge, or which seems less likely so to contribute, deserves to be dismissed or downplayed. Piecemeal scepticism about specific items of knowledge can contribute to improving knowledge by alerting us to error in what we take to be knowledge. But wholesale scepticism about great swathes of what we take to be knowledge may be inherently counterproductive, inherently incapable of contributing to the improvement of knowledge, because if what is doubted is false we may be deprived of the very means to improve knowledge: such scepticism deserves to be dismissed because of its inherently counterproductive character. Thus scepticism about specific items of apparent perceptual knowledge and knowledge of the past may well contribute to the improvement of knowledge, because the items in question may turn out, on examination, to be false. But wholesale scepticism about all perceptual knowledge, and all knowledge of the past, cannot contribute to the improvement of knowledge, and thus deserve to be dismissed on that account. If *all* apparent knowledge we acquire about objects around us via perception is false, we are deprived of the means for improving knowledge; and likewise for knowledge of the past.

AOE emerges when this modified Popperian response to scepticism is adopted in connection with science. The level 7 thesis of AOE deserves to be accepted because its acceptance can only aid, and cannot impede, the improvement of knowledge. Rivals to this thesis deserve to be dismissed because, if true, it would be impossible to acquire knowledge. The level 6 thesis of AOE deserves to be accepted because its truth implies that AOE can be implemented with hope of success, whereas its falsity implies either (i) that we cannot improve existing methods for the improvement of knowledge, or (ii) that implementing AOE will meet with apparent success for a time but will then abruptly break down in a way we could not have discovered before the breakdown occurs. Accepting that thesis 6 is false, in other words, contributes nothing to the improvement of knowledge, whereas accepting that it is true may well contribute much. Theses at levels 5 to 3 deserve to be accepted because these theses have either led to greater empirical success at levels 2 and 1 than any rival theses compatible with theses 7 and 6, or promise to do so. As I have already argued, a rival, B*, to the current level 3 thesis, B, deserves to be accepted in preference to B if (a) physical theory that accords with B* (and clashes with B) is more empirically successful than physical theory that accords with B, and (b) B* accords better with the level 4 thesis of physicalism than B does. Likewise, a rival, P*, to physicalism, P, at level 4, deserves to be accepted in preference to P if (a) physical theory that accords with P* (and clashes with P) is more empirically successful than physical theory that accords with P, and (b) P* accords with the level 5 thesis of comprehensibility at least as well as P does. In both cases, that thesis is accepted which has the greatest empirical success associated with it, and accords the best with theses above it in the hierarchy of theses. In both cases, rival theses (B* and P*) satisfying (a) and (b) would be major contributions to scientific knowledge.

The metaphysical theses of AOE are displayed in a hierarchy so that criticism can be concentrated where it seems most likely to be fruitful, namely low down in the hierarchy, and so that it may be of a type that seems most likely to be fruitful, namely leading to modifications of theses low-down in the hierarchy compatible with theses higher up.

There is no argument for the *truth*, or *probable* truth, of theses at levels 3 to 7. Rather, there are arguments to the effect that, accepting the theses at levels 3 to 7, as a part of theoretical scientific knowledge, gives us our best hope of improving knowledge of truth at levels 1 and 2.

In order fully to justify acceptance of theses at levels 3 to 5 of AOE one would need to look at various possible rival theses at these levels, and consider how the history of physics up to the present favours acceptance the theses I have specified over any of the rivals. I have in this book made remarks in this direction: see in particular the discussion of problems (2) and (5) of section 1.3 of chapter one, sections 2.1 to 2.3 of chapter two, and sections 1 to 4 above of this appendix. Elsewhere

(Maxwell, 1998, chs. 3-5) I have pursued some of these matters in more detail.

6.5 Closing the epistemological gulf between empirical and theoretical knowledge

The justificational problems of induction, (2) and (3), arise because of the epistemological gulf that seems to separate statements of observational and experimental results on the one hand, and statements of physical laws and theories on the other. "This piece of copper wire connects up these two terminals" (the kind of statement that is implicit in an account of an observation or experiment in physics) is a statement about a particular object or state of affairs, here and now, which can it seems be checked straightforwardly by observations and trials made here and now (although radical scepticism about perception seeks to undermine confidence in such assertions) . "All objects attract each other in accordance with $F = GM_1M_2/d^2$" makes a precise assertion about all objects, everywhere, at all times and places. Given the precision and universal scope of this assertion, all the evidence we might gather in support of it can only be infinitesimal and highly biased (in that it is all gathered in the minute region of space and time that we occupy), and thus hopelessly inadequate. The epistemological gulf separating evidence and theory seems unbridgeable. This is the problem of induction in its classical form. Anything which manages somehow to lessen or bridge this gulf might help solve the problem. The hope would be that somehow, against logic, theoretical knowledge could be brought closer to the security of evidential knowledge. But an argument that does the reverse, and reveals that evidential knowledge is nearly as insecure as theoretical knowledge, and for much the same reason, would also serve to lessen the epistemological gulf. Such an argument is the following.

One way of highlighting the existence of the epistemological gulf is to draw attention to the fact that whereas knowledge of physical theory presupposes metaphysical knowledge of the entire cosmos, knowledge of evidence does nothing of the kind. If we knew for certain that physicalism is true, we would have far greater grounds for being confident that the repeatedly confirmed predictions of a theory that has met with great empirical success (such as Newtonian theory) will continue to be correct, within established limits of accuracy. It is not knowing that physicalism is true that renders standard predictions of Newtonian theory, let us say, so apparently wildly uncertain and insecure. Theoretical knowledge, in other words, is so insecure because of the insecurity of presupposed metaphysical, cosmological theses; empirical knowledge is much more secure because it does not make such cosmological presuppositions.

But this latter is false. Even humble particular statements about our immediate surroundings contain presuppositions about the entire cosmos. "I can walk across the room" presupposes that nowhere in the entire universe is an explosion even now occurring of unprecedented force which will spread with nearly infinite speed to engulf the room before I can take a step. "This piece of copper wire will continue to behave as copper wire for the next few minutes" — the kind of statement any physical experiment needs to be true — presupposes that nowhere in the entire cosmos is there a new process occurring which will spread invisibly, with near infinite speed, to subvert properties of matter, so that copper ceases to behave as copper. All evidential knowledge of experimental set-ups and results in physics is of this type, and thus makes presuppositions about the entire cosmos. Cosmological presuppositions are made by *both* evidential and theoretical knowledge: the epistemological gap between the two is in reality not nearly as wide as the traditional way of posing the Humean problem of induction would lead one to believe.[32]

6.6 Can aim-oriented empiricism solve the practical problem of induction?

According to AOE, all our knowledge is, in the end, conjectural in

---

32. Some philosophers of science have argued that evidence is "theory-laden", so that universal laws are presupposed by experiments. In so far as this is true, the epistemological gap between evidence and theory disappears altogether. In principle, however, experiments do not need to presuppose laws. An experiment that includes a piece of copper wire may need to presuppose that the wire continues to behave as copper during the period of the experiment, but does not need to presuppose that all copper, everywhere, behaves as copper, at all times and places. Others argue that an experimental result in physics, in order to be capable of corroborating or refuting a theory, must be a repeatable effect, a low-level empirical hypothesis: see, for example (Popper, 1959, pp. 86-91). This, too, annihilates the gulf between experiment and theory (or perhaps places the problem of induction entirely within the experimental realm).

character. If that is the case, what basis can there be for holding that AOE solves the *practical* problem of induction, problem (3)? As I have already remarked, every time we fly in an aeroplane, cross a suspension bridge or take medicine, we entrust our lives to the correctness of the predictions of scientific theories. But these theories are, according to AOE, mere *conjectures*. How can we entrust our life to conjectures?

John Worrall has dramatized the problem as follows. We are, let us suppose, standing on top of the Eiffel Tower, and we are confronted by two rival conjectures: one says if we jump we will float gently down to earth without harm; the other says we will fall in the usual way to our death (Worrall, 1989). Only lunatics think the first a viable possibility; the rest of us are absolutely confident in the truth of the second conjecture. How is this confidence to be justified? No version of SE comes up with an adequate answer, especially as *ad hoc* versions of Newtonian theory or general relativity can be concocted which predict jumping on this occasion will lead to a soft, harmless landing, and which are empirically *more* successful than the non-*ad hoc* versions of these theories. Can AOE justify our confidence that if we jump we will be killed?

I have already pointed out that much of our (pre-Humean) common sense confidence that we will be killed if we jump may not be justifiable. What we require, here, is for AOE to justify the rationality of deciding we will be killed, even if this falls short of justifying common sense confidence in the correctness of this decision.

If we grant the truth of the theses of AOE, from level 4 to 7, a straightforward answer can be given. Physicalism tells us that a unified pattern of physical law governs all phenomena. By far our best efforts at discovering invariant (or unified) laws governing such things as bodies in free fall near the earth's surface are Newton's theory of gravitation and, better still, Einstein's. No rival theory is even remotely as good at complying with the two requirements of (a) empirical success and (b) compatibility with physicalism. Theories that are empirically more successful and predict a gentle landing can be concocted, but these clash horribly with physicalism, and deserve to be rejected for that reason. But Newton's or Einstein's theory (plus additional information about such things as the mass of the earth) predict with stark clarity: jumping leads to rapid acceleration at roughly 32 ft per sec per sec. Above all, a theory which accords with physicalism as well as Newton's or Einstein's theory, but predicts that jumping will lead to a gentle floating to the ground is nowhere on the horizon. Thus, given the truth of physicalism, there is absolutely no question, no grounds for serious doubt, whatsoever: jumping is for suicides only.

But we are not given the truth of physicalism. At most arguments deployed above give grounds for accepting physicalism granted our aim is to improve our conjectural knowledge of truth. There are arguments justifying *acceptance* of theses at levels 3 to 7 granted our aim is to improve knowledge of truth, but no arguments justifying the *truth* of these theses. And it is the latter we require, it seems, to solve Worrall's problem, and the practical problem of induction more generally.

I have two replies to this objection.

First, even in the absence of any kind of justification of the truth of physicalism, AOE succeeds, nevertheless, in distinguishing decisively between those conjectures we happen to be confident are true, to the extent even of entrusting our lives to their truth, and those conjectures (even empirically more successful conjectures) about whose truth we have no such confidence.

Second, the demand that the truth of physicalism must receive some kind of justification before it becomes rationally acceptable for practical purposes is not just impossible to fulfil; it deserves to be rejected in that it stems from an unrigorous, untenable conception of science, and human knowledge more generally. If, and only if, some version of SE is correct — and science is based on evidence, and on metaphysical theses whose truth has been justified (if there are any) — is the demand to justify the truth of physicalism itself justifiable. But all versions of SE are unrigorous and untenable. Hence, the SE demand to justify physicalism is itself unjustifiable, and must be rejected. What I have shown in section 6.5 and elsewhere is that *all* significant factual knowledge, common sense and scientific, implies (and thus presupposes) cosmological theses: rigour requires that these unjustifiable cosmological theses are made explicit, so that they can be critically assessed and, we may hope, improved. To demand that such cosmological theses cannot be accepted unless their truth is justified condemns science to lack of rigour, because it ensures that unjustifiable cosmological theses will not be, and cannot be, accepted as a part of scientific knowledge. The demand deserves to be rejected.

Human knowledge has always had this inescapable cosmological dimension built into it. The illusion that science could dispense with such unjustifiable cosmological conjectures only crept in with the general acceptance of SE, some time after Newton and before the end of the 19th century. What needs to be done is not to justify the truth of physicalism, but rather to justify the claim that this cosmological conjecture has played a more fruitful role in the advance of science than any rival at that level. I have shown how this can be done. *Science does not prove its cosmological conjectures; it sets out to improve those it has inherited from the past.* Physicalism is the best available, at that level of generality, and that suffices to solve the Worrall problem, and the justificational problem of induction (in so far as it can be solved). We are justified in entrusting our lives to the standard empirical predictions of those theories (a) which have met with sufficient empirical success, and (b) which, together with other such empirically successful theories, are more nearly compatible with our best metaphysical theses concerning the comprehensibility and knowability of the universe. Our best metaphysical theses, in turn, are those which have generated the most empirically progressive scientific research programmes. The circularity that seems to be involved here is legitimised by acceptance of meta-knowability.

The problem of induction is not just a philosophical puzzle *à la* Wittgenstein. Its long-standing insolubility is indicative of a fundamental defect in our understanding of science and its relationship with metaphysics and philosophy — a fundamental defect in our whole culture. Science does not stand opposed to metaphysics and philosophy; it is metaphysics and philosophy carried on by other means, employing the improved methods of investigation of empiricism: observation and, above all, experimentation (a point enshrined in the 17th century terms of "experimental" and "natural" philosophy). A basic task for philosophers today is to try to get across to the scientific community just how vital metaphysics and philosophy, properly conducted, are for science, so that scientists and philosophers can begin to collaborate on implementing AOE science, thus recreating natural philosophy, and curing science of some of its neurosis. And this needs to be part of a broader campaign to encourage academics in general to start implementing wisdom-inquiry, so that academia may be cured of some of its neurosis, and may begin to help humanity learn how to solve its conflicts and problems of living in increasingly cooperatively rational ways.

References

Abott, P. and Wallace, C. (1990). *An Introduction to Sociology: Feminist Perspectives*. London: Routledge.

Aitchison, I. J. R. (1985). Nothing's Plenty: The Vacuum in Modern Quantum Field Theory, *Contemporary Physics 26*: pp. 333-391.

Aitchison, J. J. R. and A. J. G. Hey (1982). *Gauge Theories in Particle Physics*. Bristol: Adam Hilger.

Appleyard, B. (1992). *Understanding the Present: Science and the Soul of Modern Man*. London: Picador.

Armstrong, D. M. (1968). *A Materialist Theory of the Mind*. London: Routledge and Kegan Paul.

Atkins, P. W. (1983). *Molecular Quantum Mechanics*, Oxford: Oxford University Press.

Austin, J. L. (1962). *Sense and Sensibilia*. Oxford: Oxford University Press.

Ayer, A. J. (1936). *Language, Truth and Logic*. London: Gollancz.

Barnes, B. (1974). *Scientific Knowledge and Sociological Theory*. London: Routledge and Kegan Paul.

Barnes, B. and Bloor, D. (1981). Relativism, Rationalism and the Sociology of Knowledge: in M. Hollis and S. Lukes (eds.) *Rationality and Relativism*. Oxford: Blackwell, pp. 21-47.

Barnes, B., D. Bloor and J. Henry (1996). *Scientific Knowledge: A Sociological Analysis*. University of Chicago Press, Chicago.

Barrett, W. (1962). *Irrational Man*. New York: Doubleday.

Barrow, J. (1988). *The World within the World*. Oxford: Clarendon Press.

Bell, J. S. (1987). *Speakable and Unspeakable in Quantum Mechanics*. Cambridge: Cambridge University Press.

Berlin, I. (1979). *Against the Current*. London: Hogarth Press.

———— (1999). *The Roots of Romanticism*. London: Chatto and Windus.

Berman, M. (1981). *The Reenchantment of the World*. Ithaca: Cornell University Press.

Bloor, D. (1991). *Knowledge and Social Imagery*, 2nd edn., University of Chicago Press, Chicago.

Boscovich, R. J. (1966). *A Theory of Natural Philosophy*, translated by J. M. Child. Cambridge, Mass.: MIT Press (first published in 1763).

Brown, J. R. (2001). *Who Rules in Science?*. Cambridge, Mass.: Harvard University Press.

Buchwald, J. (1985). *From Maxwell to Microphysics*. Chicago: Chicago University Press.

Burtt, E. A. (1932). *The Metaphysical Foundations of Modern Physical Science*. London: Routledge and Kegan Paul.

Carnap, R. (1950). *Logical Foundations of Probability*. London: Routledge and Kegan Paul.

Carson, R. (1965). *Silent Spring*. Harmondsworth: Penguin (first published 1962).

Chalmers, D. (1996). *The Conscious Mind*. Oxford: Oxford University Press.

Club of Rome, (1974). *The Limits to Growth*. London: Pan (first published 1972).

Commoner, B. (1972). *The Closing Circle*. New York: Bantam Books (first published 1971).

Cotes, R. (1962). Cotes's Preface to the Second Edition, in I. Newton, *Principia*. Berkeley: University of California Press, xx-xxxiii (first published 1713).

Damasio, A. (1994). *Descartes' Error*. New York: Putnam.

_____ (2000). *The Feeling of What Happens*. London: Vintage.

Davies, P. C. W., and Brown, J. (eds.), (1988). *Superstrings: A Theory of Everything?*. Cambridge: Cambridge University Press.

Dennett, D. (1966). *Darwin's Dangerous Idea*. London: Penguin Books.

Duhem. P. (1954). *The Aim and Structure of Physical Theory*. Princeton: Princeton University Press (first published in French in 1906).

Dupré, J. (1993). *The Disorder of Things*. Cambridge, Mass.: Harvard University Press.

Durant, J. (1997). Beyond the Scope of Science, *Science and Public Affairs*, Spring, pp. 56-57.

Earman, J. (1992). *Bayes or Bust? A Critical Examination of Bayesian Confirmation Theory*. Cambridge, Mass.: MIT Press.

Einstein, A. (1949). Autobiographical Notes, in P. A. Schilpp (ed.), *Albert Einstein: Philosopher-Scientist*. La Salle, Illinois: Open Court, pp. 1-94.

Fargaus, J. (ed.) (1993). *Readings in Social Theory*. New York: McGraw-Hill.

Feyerabend, P. (1978). *Against Method*. London: Verso.

_____ (1987). *Farewell to Reason*. London: Verso.

Feynman, R., R. Leighton and M. Sands (1965), *The Feynman Lectures on Physics vol. II*. Reading, Mass.: Addison-Wesley.

Fox Keller, E. (1984). *Reflections on Gender and Science*. New Haven: Yale University Press.

Freud, S. (1962). *Two Short Accounts of Psycho-Analysis*. Harmondsworth: Penguin.

Friedman, M. (1974), Explanation and Scientific Understanding, *Journal of Philosophy 71*: pp.5-19.

Fuller, S. (1993). *Philosophy of Science and Its Discontents*. New York: Guildford Press.

Gascardi, A. (1999). *Consequences of Enlightenment*, Cambridge: Cambridge University Press.

Gay, P. (1973). *The Enlightenment: An Interpretation*. London: Wildwood House.

Goodman, N. (1954). *Fact, Fiction and Forecast*. London: Athlone Press.

_____ (1972), *Problems and Projects*. New York: Bobbs-Merrill.

Greenburg, D. S. (1971). *The Politics of Pure Science*. New York: New American Library.

Greene, B. (1999). *The Elegant Universe*. New York: Norton.

Griffiths, D. (1987). *Introduction to Elementary Particles*. New York: John Wiley.

Gross, P. and N. Levitt (1994). *Higher Superstition: The Academic Left and It Quarrels with Science*. Baltimore: John Hopkins University Press.

Gross, P., N. Levitt and M. Lewis (eds.) (1996). *The Flight from Science and Reason*. Baltimore: John Hopkins University Press.

Grünbaum, A. (1984). *The Foundations of Psychoanalysis*. Berkeley: University of California Press.

Hacohen, M. H. (2000). *Karl Popper — The Formative Years, 1902-1945.* New York: Cambridge University Press.

Harding, S. (1986). *The Feminist Question in Science.* Milton Keynes: Open University Press.

Harman, G. (1965). The Inference to the Best Explanation, *Philosophical Review 74,* pp. 88-95.

_____ (1968). Enumerative Induction as Inference to the Best Explanation, *Journal of Philosophy 64,* pp. 529-533.

Hayek, F. A. (1979). *The Counter-Revolution of Science.* Indianapolis: LibertyPress.

Hempel, C. G. ((1965). *Aspects of Scientific Explanation.* New York: Free Press.

Hesse, M. (1965). *Forces and Fields.* Totowa, New Jersey: Littlefield, Adams and Co.

_____ (1974). *The Structure of Scientific Inference.* London: Macmillan.

Higgins, R., (1978). *The Seventh Enemy.* London: Hodder and Stoughton.

Holton, G. (1973). *Thematic Origins of Scientific Thought.* Cambridge, Mass.: Harvard University Press.

Howson, C. (2000). *Hume's Problem.* Oxford: Oxford University Press.

Howson, C. and Urbach, P. (1993). *Scientific Reasoning: The Bayesian Approach.* Chicago: Open Court.

Hunt, B. J. (1991). *The Maxwellians.* New York: Cornell University Press.

Hume, D. (1959). *A Treatise of Human Nature, vol. 1.* London: Dent (first published 1739).

Isham, C. (1989). *Lectures on Groups and Vector Spaces for Physicists.* London: World Scientific.

_____ (1997). Structural Issues in Quantum Gravity, *General Relativity and Gravity: GR 14.* Singapore: World Scientific, pp. 167-209.

Jeffreys, H. and D. Wrinch (1921), On Certain Fundamental Principles of Scientific Enquiry, *Philosophical Magazine 42,* pp. 269-298.

Jones, H. (1990), *Groups, Representations and Physics.* Bristol: Adam Hilger.

Kane, R. (1996). *The Significance of Free Will.* Oxford: Oxford University Press.

Kant, I. (1961). *Critique of Pure Reason.* London: Macmillan (first published in German in 1781).

Kitcher, P. (1981), Explanatory Unification, *Philosophy of Science 48,* pp. 507-531.

_____ (1989). Explanatory Unification and Causal Structure, in P. Kitcher and W. C. Salmon (eds.) *Scientific Explanation, Minnesota Studies in the Philosophy of Science, xiii.* Minneapolis: University of Minnesota Press, pp. 428-448.

_____ (2001). *Science, Truth and Democracy.* Oxford: Oxford University Press.

Koertge, N. (ed.) (1998). *A House Built on Sand.* Oxford: Oxford University Press.

Koyré, A. (1965). *Newtonian Studies.* London: Chapman and Hall.

Kuhn, T. S. (1970). *The Structure of Scientific Revolutions.* Chicago: Chicago University Press (first published in 1962).

Kukla, A. (2001). Theoreticity, Undetermination, and the Disregard for Bizarre Scientific Hypotheses, *Philosophy of Science 68,* pp. 21-35

Kyburg, H. (1970). *Probability and Inductive Logic.* London: Collier-Macmillan.

Lacey, H. (1999). *Is Science Value Free?.* London: Routledge.

Laing, R. D. (1965). *The Divided Self.* Harmondsworth: Penguin.

Lakatos, I. (1970). Falsification and the Methodology of Scientific Research Programmes, in I. Lakatos and A. Musgrave, (eds.) *Criticism and the Growth of Knowledge.* Cambridge: Cambridge University Press, pp. 91-196.

_____ (1976). *Proofs and Refutations.* Cambridge: Cambridge University Press.

Latour, B. (1987). *Science in Action.* Milton Keynes: Open University Press.

Laudan, L. (1977). *Progress and Its Problems.* London: Routledge and Kegan Paul.

Lenski, G. et al. (1995). *Human Societies: An Introduction to Macrosociology.* New York: McGraw-Hill.

Lipton, P. (2004). *Inference to the Best Explanation,* 2nd ed. London: Routledge.

Macionis, J. and Plummer, K. (1997). *Sociology: A Global Introduction.* New York: Prentice Hall.

Mandl, F. and G. Shaw, ((1986). *Quantum Field Theory.* Chichester: John Wiley.

Marcuse, H. (1964). *One Dimensional Man.* Boston: Beacon Press.

Maxwell, N. 1968. Can there be Necessary Connections between Successive Events?, *British Journal for the Philosophy of Science 19*, pp. 1-25.

_____ (1972). A New Look at the Quantum Mechanical Problem of Measurement, *American Journal of Physics, 40*, pp. 1431-1435.

_____ (1973a). Alpha Particle Emission and the Orthodox Interpretation of Quantum Mechanics, *Physics Letters 43A*, pp. 29-30.

_____ (1973b). The Problem of Measurement — Real or Imaginary?", *American Journal of Physics 41*, pp.1022-1025.

_____ (1974). The Rationality of Scientific Discovery, *Philosophy of Science 41*, pp. 123-153 and 247-295.

_____ (1975). Does the Minimal Statistical Interpretation of Quantum Mechanics Resolve the Measurement Problem?, *Methodology and Science 8*, pp. 84-101.

_____ (1976a). *What's Wrong With Science?.* Frome: Bran's Head Books.

_____ (1976b). Towards a Micro Realistic Version of Quantum Mechanics, *Foundations of Physics 6*, pp. 275-292 and 661-676.

_____ (1977). Articulating the Aims of Science, *Nature 265*, p. 2.

_____ (1979). Induction, Simplicity and Scientific Progress, *Scientia 114*, pp. 629-653.

_____ (1980). Science, Reason, Knowledge and Wisdom: A Critique of Specialism, *Inquiry 23*, pp. 19-81.

_____ (1982). Instead of Particles and Fields: A Micro Realistic Quantum 'Smearon' Theory, *Foundations of Physics 12*, pp.607-631.

_____ (1984a). *From Knowledge to Wisdom: A Revolution in the Aims and Methods of Science.* Oxford: Basil Blackwell.

_____ (1984b). From Knowledge to Wisdom: Guiding Choices in Scientific Research. Delivered as a lecture by invitation to the Annual Meeting of the AAAS, New York, May, 1984, and published in *Bulletin of Science, Technology and Society 4*, pp. 316-334.

_____ (1985a). From Knowledge to Wisdom: the Need for an Intellectual Revolution", *Science, Technology and Society Newsletter 21*, pp. 55-63.

_____ (1985b). Are Probabilism and Special Relativity Incompatible?, *Philosophy of Science 52*, pp. 23-44.

_____ (1987). Wanted: a new way of thinking, *New Scientist*, 14 May, p. 63.

_____ (1988). Quantum Propensiton Theory: A Testable Resolution of the Wave/Particle Dilemma, *British Journal for the Philosophy of Science 39*, pp. 1-50.

_____ (1991). How Can We Build a Better World?, *Einheit der Wissenschaften: Internationales Kolloquium der Akademie der Wissenschaften zu Berlin, 25-27 June 1990*, ed. J. Mittelstrass. Berlin and New York: Walter de Gruyter, pp. 388-427.

_____ (1992a). What Kind of Inquiry Can Best Help Us Create a Good World?, *Science, Technology, and Human Values 17*, pp. 205-227.

_____ (1992b). What the Task of Creating Civilization has to Learn from the Success of Modern Science: Towards a New Enlightenment, *Reflections on Higher Education 4*, pp. 47-69.

_____ (1993a). Induction and Scientific Realism: Einstein versus van Fraassen, *British Journal for the Philosophy of Science 44*, pp. 61-79, 81-101 and 275-305.

_____ (1993b). Beyond Fapp: Three Approaches to Improving Orthodox Quantum Theory and An Experimental Test, in *Bell's Theorem and the Foundations of Modern Physics*, edited by A. van der Merwe, F. Selleri and G. Tarozzi. Singapore: World Scientific, pp. 362-370.

_____ (1993c). Does Orthodox Quantum Theory Undermine, or Support, Scientific Realism?, *The Philosophical Quarterly 43*, pp. 139-157.

_____ (1994a). Towards a New Enlightenment: What the Task of Creating Civilization has to learn from the Success of Modern Science, in *Academic Community: Discourse or Discord?*, edited by R. Barnett. London: Jessica Kingsley, pp. 86-105.

_____ (1994b). Particle Creation as the Quantum Condition for Probabilistic Events to Occur, *Physics Letters A 187*, pp. 351-355.

_____ (1997a). Science and the Environment: A New Enlightenment, *Science and Public Affairs*, Spring 1997, pp. 50-56.

_____ (1997b). Must Science Make Cosmological Assumptions if it is to be Rational?, in *The Philosophy of Science: Proceedings of the Irish Philosophical Society Spring Conference*, edited by T. Kelly. Maynooth: Irish Philosophical Society, pp. 98-146.

_____ (1998). *The Comprehensibility of the Universe: A New Conception of Science*. Oxford: Oxford University Press.

_____ (1999). Has Science Established that the Universe is Comprehensible?, *Cogito 13*, pp. 139-145.

_____ (2000a). Can Humanity Learn to become Civilized? The Crisis of Science without Civilization, *Journal of Applied Philosophy 17*, pp. 29-44.

_____ (2000b). A New Conception of Science, *Physics World 13*, No. 8, pp. 17-18.

_____ (2001a). Wisdom and Curiosity? I Remember them Well, *The Times Higher Education Supplement*, No. 1,488, 25 May, p. 14.

_____ (2001b). Can Humanity Learn to Create a Better World? The Crisis of Science without Wisdom, in *The Moral Universe*, edited by T. Bentley and D. Stedman Jones, *Demos Collection 16*, pp. 149-156.

_____ (2001c). *The Human World in the Physical Universe: Consciousness, Free Will and Evolution.* Lanham, Maryland: Rowman and Littlefield.

_____ (2002a). The Need for a Revolution in the Philosophy of Science, *Journal for General Philosophy of Science 33*, pp. 381-408.

_____ (2002b). Karl Popper, in *British Philosophers, 1800-2000,* edited by P. Dematteis, P. Fosl and L. McHenry, *Dictionary of Literary Biography 262,* Bruccoli Clark Layman, Detroit, pp. 176-194.

_____ (2003a). Two Great Problems of Learning, *Teaching in Higher Education 8,* pp. 129-134.

_____ (2003b). Do Philosophers Love Wisdom?, *The Philosophers' Magazine,* Issue 22, 2nd quarter, pp. 22-24.

_____ (2003c). Science, Knowledge, Wisdom and the Public Good, *Scientists for Global Responsibility Newsletter 26,* February, pp. 7-9.

_____ (2004a). Popper, Kuhn, Lakatos and Aim-Oriented Empiricism, *Philosophia 32* (forthcoming). See also http://www.philsci-archive.pitt.edu/archive/00000251/

_____ (2004b). Scientific Metaphysics, http://www.philsci-archive.pitt.edu/archive/00001674/

_____ (2004c). Does Probabilism Solve the Great Quantum Mystery?, http://www.philsci-archive.pitt.edu/archive/00001704/

_____ (2004d). Non-Empirical Requirements Scientific Theories Must Satisfy: Simplicity, Unification, Explanation, Beauty, http://www.philsci-archive.pitt.edu/archive/ 00001759/

_____ (2004e). Comprehensibility rather than Beauty, http://www.philsci-archive.pitt.edu/archive/00001770/

_____ (2004f). In Defense of Seeking Wisdom, *Metaphilosophy 35,* pp. 733-743.

McAllister, J. (1996). *Beauty and Revolution in Science.* Ithaca: Cornell University Press.

Miller, D. (1994). *Critical Rationalism.* Chicago: Open Court.

Moriyasu, K. (1983). *An Elementary Primer for Gauge Theory.* Singapore: World Scientific.

Nagel. T. (1979). *Mortal Questions.* Cambridge: Cambridge University Press.

Newton-Smith, W. (1981). *The Rationality of Science.* London: Routledge and Kegan Paul.

Nola, R. and H. Sankey (2000). A Selective Survey of Theories of Scientific Method, in R. Nola and H. Sankey (eds.) *After Popper, Kuhn and Feyerabend.* Dordrecht: Kluwer.

O'Hear, A. (1989). *An Introduction to the Philosophy of Science:* Oxford: Oxford University Press.

Peirce, C. S. (1958). *Values in a Universe of Chance,* edited by P. P. Wiener. New York: Doubleday.

Pickering, A. (1984). *Constructing Quarks.* Chicago: University of Chicago Press.

Place, U. T. (1956). Is Consciousness a Brain Process?, *British Journal of Psychology 46,* pp. 44-50.

Poincaré, H. (1952). *Science and Hypothesis.* New York: Dover.

Polanyi, M. (1958). *Personal Knowledge.* London: Routledge and Kegan Paul.

Popper, K. (1959). *The Logic of Scientific Discovery*. London: Hutchinson (first published in German in 1934).

_____ (1963). *Conjectures and Refutations*. London: Routledge and Kegan Paul.

_____ (1969). *The Open Society and Its Enemies*. London: Routledge and Kegan Paul.

_____ (1972). *Objective Knowledge*. Oxford: Oxford University Press.

_____ (1974). *The Poverty of Historicism*. London: Routledge and Kegan Paul.

_____ (1976). *Unended Quest*. Glasgow: Fontana.

Quine, W. V. O. (1961). Two Dogmas of Empiricism. Chapter 2 of *From a Logical Point of View*. Cambridge: Harvard University Press.

Regan, T. (2004). *Animal Rights, Human Wrongs*. Lanham, Maryland: Rowman and Littlefield.

Reichenbach, H. (1961). *Experience and Prediction*, Chicago: Chicago University Press (first published 1938).

Roszak, T. (1973). *Where the Wasteland Ends*. London: Faber and Faber.

Ryle, G. (1949). *The Concept of Mind*. London: Hutchinson.

Salmon, W. ((1989), *Four Decades of Scientific Explanation*. Minneapolis: University of Minnesota Press.

Schumacher, E. F., (1973). *Small is Beautiful*. London: Blond and Briggs.

Schutz, B. F. (1989). *A First Course in General Relativity*. Cambridge: Cambridge University Press.

Schwartz, B. (1987). *The Battle for Human Nature*. New York: W. W. Norton.

Segerstrale, U. (ed.) (2000). *Beyond the Science Wars*. Albany: State University of New York Press.

Shapin, S. (1994). *A Social History of Truth*. Chicago: University of Chicago Press.

Shapin, S. and S. Schaffer (1985). *Leviathan and the Airpump*. Princeton: Princeton University Press.

Singer, P. (1995). *Animal Liberation*. London: Pimlico.

_____ (2002). *One World: The Ethics of Globalization*. New Haven: Yale University Press.

Smart, J. J. C. (1963). *Philosophy and Scientific Realism*. London: Routledge and Kegan Paul.

Smolin, L. (2000). *Three Roads to Quantum Gravity*. London: Weidenfeld and Nicolson.

Snow, C. P. (1964). *The Two Cultures and a Second Look*. Cambridge: Cambridge University Press.

Sober, E. (1975), *Simplicity*. Oxford: Oxford University Press.

Sokal, A. and Bricmont, J. (1998). *Intellectual Impostures*. London: Profile Books.

Swain, M. (ed.) (1970). *Induction, Acceptance and Rational Belief*. Dordrecht: D. Reidel.

Tischler, H. (1996). *Introduction to Sociology*. Orlando: Harcourt Brace.

van Fraassen, B. (1980). *The Scientific Image*. Oxford: Clarendon Press.

_____ (1985). Empiricism and the Philosophy of Science, in P. M. Churchland and C. A. Hooker (eds.) *Images of Science*. Chicago: University of Chicago Press, pp. 245-308.

Watkins, J. W. N. (1984). *Science and Scepticism.* Princeton: Princeton University Press.

Weinberg, S. (1993). *Dreams of a Final Theory.* London: Hutchinson.

Wittgenstein, L. (1953). *Philosophical Investigations.* Oxford: Blackwell.

Wolpert, L. (1993). *The Unnatural Nature of Science.* London: Faber and Faber.

Worrall, J. (1989). Structural Realism: The Best of Both Worlds?, *Dialectica 43*, pp. 99-124.

Worrall, J. (1989). Why Both Popper and Watkins Fail to Solve the Problem of Induction, in *Freedom and Rationality: Essays in Honour of John Watkins.* Dordrecht: Kluwer.

Yang, C. N. and R. L. Mills (1954). *Physical Review 96*, p. 191.

Zamyatin, Y. (1972). *We.* Harmondsworth: Pengin.

Ziman, J. (1968). *Public Knowledge.* Cambridge: Cambridge University Press.

Index

Abott, P., 118
academic inquiry
  and apathy, 141-143
  curing neurosis of, 74, 82-83, 118-121,
  129, 145-147
  damaging consequences of irrationality
  of, 93-94
  dominated by knowledge-inquiry,
  117-118
  and education, 94, 103, 117, 135,
  136-140, 145
  and the Enlightenment, 72-74
  and government, 88, 113
  inefficacy of, 144-147
  irrationality of, 72, 73, 84-85, 92,
  115-116, 123-124
  as mixture of knowledge- and wisdom-
  inquiry, 117
  need for revolution of, xiv-xv, 82-83,
  118-121
  neurosis of, 72, 74, 113, 129
  and problems of living, 92, 93, 117
  questions concerning neurosis of, 113
  relationship with social world, 120, 145
  and Romanticism, 73
  and Traditional Enlightenment, 72-74,
  80-81
  wisdom as proper aim for, 81, 82,
  88-89, 103
action before knowledge, 123-124
Adorno, T. W., xii
agriculture
  and the Enlightenment programme, 81
  and science, 52, 70, 85
  sustainable, 96
  and wisdom-inquiry, 86, 96-97, 101,
  117, 133, 134
AIDS epidemic, 132, 134
aim-oriented empiricism (AOE), 19-28,
  175-178

arguments for, 18-22, 24-28, 153-159,
  176-191
circularity objection and its solution,
  207-210
and diverse branches of science, 41-47
exposition of, 19-24, 175-178
and knowledge of truth, 179-180
level 2 thesis of, 22, 175
level 3 thesis of, 22-23, 175, 182-183
level 4 thesis, 23, 175, 182-185
level 5 thesis of, 23, 175-176, 181-182,
  183-184
level 6 thesis of, 23, 176, 177, 180-181,
  184, 208-210
level 7 thesis of, 23, 26, 176, 180
and natural philosophy, 47-48
neglect of, 151-153
and politics, 66
and Popper, 79-80, 177-178, 213-215
and rationally discoverable, 23, 177,
  181, 184
and rational scientific discovery, 34-39
rigour of, 24, 25n, 178-179
rival theses to, 182-191,
and scientific method, 44-47
and scientific revolutions, 27-28, 37-39
and search for explanation and
  understanding, 49-50
and simplicity of theory, 172-174
and solution to problem of induction,
  205-220
and unity of theory, 155, 158-159,
  160-174
and values, 58-60
see also comprehensibility of universe;
  physical comprehensibility, thesis of
aim-oriented rationality, 22
  and civilization, 95-97
  and cooperative conflict resolution,
  97-98